The Fruits of Natural Advantage

The Fruits of Natural Advantage

Making the Industrial Countryside in California

Steven Stoll

UNIVERSITY OF CALIFORNIA PRESS

Berkeley / Los Angeles / London

University of California Press
Berkeley and Los Angeles, California

University of California Press, Ltd.
London, England

© 1998 by
The Regents of the University of California

Library of Congress Cataloging-in-Publication Data

Stoll, Steven, 1966–
    The fruits of natural advantage : making the industrial
countryside in California / Steven Stoll.
        p.   cm.
    Includes bibliographical references and index.
    ISBN 0–520–21172–3 (cloth : alk. paper)
    1. Fruit industry—California.   2. Fruit—California—Marketing.
3. Horticulture—California.   4. Agriculture—Economic aspects—
California.   I. Title.
HD9247.C2S73   1998
338.1′74′09794—dc21   97–27015
                                                                        CIP

Portions of chapter 4 are reprinted from Steven Stoll, "Insects and
Institutions: University Science and the Fruit Business in California,"
Agricultural History 69 (spring 1995): 216–39, by permission.

Printed in the United States of America
9   8   7   6   5   4   3   2   1

For Sara
Song of Songs 6–7

*The real farmer is said to adhere too closely
to the ways of his father. He plods. The city
man who goes to the country would correct
all this by overturning it. He sees revolution
in everything.*

L. H. Bailey,
"The Collapse of Freak Farming" (1903)

*As population increases, as the prices of
agricultural products advance, as agriculture becomes more intensive and commercialized, the use of the land must be fitted to
the geographic conditions with greater care
and precision.*

Oliver E. Baker, "The Increasing Importance
of the Physical Conditions in Determining the
Utilization of Land for Agriculture" (1921)

*The climate, seasons and soil of California
differ so materially from those of other regions, in almost every particular, and are
so diverse, even in different parts of the
State, that in many operations we are compelled to deviate from old established rules,
and frame a system of our own.*

California Horticulturist and Floral Magazine
(November 1870)

Contents

Illustrations

Preface

I wrote this book to teach myself something about agriculture and to revisit a few mysterious relics from my childhood. Like any other California kid, I saw the most familiar buildings and open spaces fold up into dirt lots, soon to be replaced by tedious fabrications of frame and stucco—never as interesting or as beautiful as what was there before. There came a time, while I was still young, when I realized that this furious trend called "development" had been under way for many years and that, not long before, the place where I lived had looked very different than it did to me. Alfalfa and orange trees once claimed the soil under the sidewalk in the neighborhood where I grew up, but not a patch of cultivated land remained in the 1970s. Always thrown for a loss by places neglected and things weathered, I learned to keep an eye out for fragments from the past. On travels to the suburban edge I caught sight of fields and orchards and paid close attention. Though I gave no particular meaning to that inland space between Los Angeles and San Francisco or to the irrigated desert near Riverside and San Bernardino, what I saw of these places put me into a creative confusion about my place in the world: trees in patterns, smudge pots on the roadside, pickers in hats with sacks and trays, packinghouses with corrugated steel roofs. I'm still searching for something in these places—still thrown for a loss by fragments in the landscape—though I ask different questions about them now.

Perhaps I am impressed with such things because I am not from rural folks. No one in my family owns a farm (although I have a wayward cousin in Vermont who raises organic vegetables); in fact, I am reasonably sure that if I could trace my family lines back three centuries, sorting through a lineage of merchants and maybe a few rabbis, there would not be a peasant farmer in all of Germany or the Russian Empire with one of my names. The name Neuburger, from my mother's side, translates to "new city dweller," suggesting a narrative in which my last farming relatives lost their property in the Bavarian countryside during some hateful uprising and moved to neighboring towns (like Neuburg?) where they entered urban occupations. The name they took amid the violence of those events cut their ties with the countryside and expressed their optimism for life in the medieval city. That city made them no promises, and they never returned to the countryside, although that choice is now open to me.

This book is my attempt to fathom the mysteries of the countryside. It is a story of California, its lowlands and foothills, coastal plains and valleys, and how a class of entrepreneurs used these places for the intensive cultivation of fruit. Through this narrative the reader will visit vineyards and packinghouses; meet state bureaucrats, merchant capitalists, and university scientists; and walk through wholesale markets and neighborhood groceries to better observe a natural environment being fixed into a worldwide economy. The setting is always somewhere between the soil and the shopping cart, where migrant workers cut the stems and stacked the boxes, where agents of the experiment station tested insecticides on infested orchards, where producer-owned corporations graded the perfect pear and advertised it to people who had never grown one themselves.

This is also a book about American agriculture in the twentieth century. The fruit business in California emerged during a time of public apprehension about the capacity of farmers to feed the people who lived in cities. California became the ultimate city-serving countryside, and its rise marked important changes in the process and purpose of farming. When fruit growers found ways to sell the most delicate crops to consumers on the far side of North America they redefined the limits of agriculture as a commercial enterprise. They helped to establish industrial methods and assumptions in the American countryside.

A capitalistic class of farmers—fruit growers—are the central actors in the story I tell. I am concerned with their goals and ideas, the plants they selected to cultivate, the landscape they influenced, the organizations they founded, and their relationship with harvest workers. The growers stood in the middle of change, leading and following in different situations, demanding or reacting at different times, and usually causing the very problems they tried to solve. There is no way to understand industrial agriculture in California without them. Chinese, Japanese, Armenian, and Mexican growers cultivated orchards all across California, but white (European and Anglo-American) growers never invited them to the annual conventions to voice their opinions. Since I am interested in the formation of an industry, I depend on these conventions and follow the activities of their foremost participants—the leading fruit growers in California. Because these people so often experienced the same problems and sought the same kind of help, I often refer to them as a singular entity.

Before I begin I need to define a few key terms, part of the common language that an author must establish with a reader. The following pages make frequent mention of four words: intensive, extensive, specialized, and industrial. *Intensive* cultivation is the application of capital and technology to increase yields on existing land. It is opposed to *extensive* cultivation, or the increase in yields through the application of new land with existing methods. The connection between high technology and intensive farming is not as obvious as it may seem. When Hidatsa and Mandan women on the banks of the upper Missouri River planted corn, beans, and squash in the same space, they saved moisture, replenished soil fertility, reduced damage from insects, and thus practiced a form of intensive cultivation. On the other hand, when farmers from Virginia to Illinois purchased McCormick reapers in the 1850s they used the latest industrial technology for extensive ends.

In the economic interdependence and labor relations they brought, machines certainly represented industrialism come to agriculture, but they did not represent intensive cultivation, and the difference is critical. In the hands of nineteenth-century farmers, reapers and combines only magnified the number of acres a sodbuster could bust. These inventions resulted in more land under cultivation, faster exploitation, and all kinds of other changes in work and rural life, but they did not manifest a new relationship between agriculture and land—only a more extensive one. Intensive cultivation made growers rend every possible dollar from their orchards while maintaining them in a high state of cultivation. To get the most out of valuable acres and high transportation costs, growers needed an entire series of institutions to keep the fruit business profitable. I will argue throughout this book that intensive cultivation became the very engine of industrial agriculture in California.

The second term is *specialization*. Specialization, otherwise known as monoculture or single-crop farming, is the central concept of the following narrative. Historians of agriculture and regional development often use specialization to describe a complex of crops—like corn and hogs—to which farmers adjusted their cultivation. A farm may be said to be "specialized" when a group of closely associated animals and plants makes up a significant portion of its output. A farm in the corn belt, for example, might have a few dairy cows, a chicken coop, a field of oats, and a kitchen garden and still derive most of its income from corn and hogs.[1] Specialization in California meant something else. As a region, California looked like one big diversified farm—almost every fruit (except bananas), vegetable, grain, and domestic animal could be found on a farm somewhere in the state. Individual orchards, however, tended to grow one or only a few varieties with no grains, no animals (a few horses before the automobile and perhaps a dairy cow), and no kitchen garden (this varied with place and time).

Technical though this may sound, the single crop is an enormous event in the history of the North American environment. Intensive specialization meant the cultivation of certain crops to the exclusion of all others. The expense of production and marketing required growers to use the natural environment to maximum effect, often leading them to put in only those crops that would yield in a particular locality better than in any other. The single crop represented a new conception of arable land, a new relationship between farmers and the people who lived in cities, a new intensity in cultivation. Once growers installed the single crop in the landscape, they realized its many insecurities and moved to protect it with all the scientific, commercial, and political power that they could gather. Specialization, along with the convulsions it caused, is the subject of this book.

Specialization implies our fourth key word: *industrialization*. The single crop represents a division of labor and also a high degree of interdependence between regional economies, business firms, government agencies, and individuals—all characteristics of industrial production. Most of all, specialized crops illustrate the separation between production and consumption, the complete commercialization of farming.[2] They have no other purpose but to feed the greater economy and the balance of foreign trade. I also take *industrial* to mean a routinized process, bound by expensive inputs, invested with capital, and operated by workers. In yet another sense, fruit growing became an *industry*, meaning that it joined a complex of institutions all engaged in different facets of the same product. Just as automobiles required glass, rubber, steel, and assembly, fruit called for pesticides, marketing networks, government regulation, and labor. These things became essential to the production and sale of fruit and brought the California countryside deep into the sphere of American capitalism.

Finally, I call fruit growing *horticulture*, or the cultivation of flowering plants. The term ordinarily describes the work of gardens, but it serves to distinguish fruits and truck crops from wheat and corn. My apologies to gardeners who would like to keep the word sheltered from the din of big business. I employ the word from time to time, though I also use *agriculture* as a general term that includes fruit growing (its strict definition is the cultivation of fields). Growers often called themselves "horticulturists" to distinguish themselves from other farmers. I use the term, along with *vineyardist* and *orchardist*, simply to add variety to my prose.

Before I enter into the elements of the argument, I should inform the reader of the subjects that this book does not treat. The *Fruits of Natural Advantage* is not a social history of rural life in California, nor does it detail the formation of towns or the day-to-day passages of rural life. I have little to say about two other subjects often associated with the industrial countryside in the Far West: irrigation and railroads.

Fruit is mostly water, so the delivery of moisture to places where rain seldom fell made the business possible. Water influenced where people settled to farm and it caused high land prices. Water is that thing without which there could be nothing. That said, the subject of water is unimportant to the events I describe. Indeed, after landowners and private water companies established a system of canals and ditches beginning in the 1880s, and after the legal battles over how rivers would be apportioned reached a fragile resolution late in the same decade, growers rarely worried out loud about irrigation until the 1920s.

I will make the same argument for why I pay almost no attention to railroads. Though without them there could not have been a commercial agriculture on the scale that growers contemplated, outside of an occasional resolution for lower rates and faster service, and after refrigeration became functional over long distances, transportation did not command debate when fruit growers assembled. As in the case of irrigation, I simply assume the railroad's importance and concentrate on other subjects. My subject is agriculture itself, defined as a set of practices to manipulate plants in environments, and how a certain manner of cultivation enabled growers in California to join a distant economy.

Finally, I consider my subject to include irrigated fruit-growing wherever it appeared, following the assumption that the similarities between localities are more important than their differences. The events that concern me happened all over California for the same reasons in each place. The central chapters of this book trace events in roughly the same period, 1880 to 1930—the fifty years between the California wheat boom and the rural depression that merged into the Great Depression, beginning with the crash in commodity prices following World War I. Each of the key institutions central to industrial farming in California had come into being by the end of this half-century.

What follows is the story of a remarkable and disturbing rural world. It is a history that is never as simple as farms turning into factories, a history in which big changes unfold from small details. Some advice for reading: Keep a close eye on the trees and vines. Notice the social and economic implications of the crop system, how patterns on the ground make other patterns. As much as this book asks questions about events in California, it also asks questions about the many ways in which economies, people, and environments meet in the act of cultivation.

Acknowledgments

I needed all kinds of help to write this book. I made several attempts to study my subject up close. I visited the orange grove and packing facilities owned and operated by Mr. T. H. Wilson and his three sons. They gave me a brief tutorial in pesticide-free horticulture and explained to me why their fifty-five acres in Riverside County turn out such excellent citrus. This brief inland excursion, in addition to trips with Sara to observe pears along the Sacramento River and raisins on the Fresno plains, sums up my experience with agriculture. I learned the details of my subject by digging in the library. And because everything came new to me, I never tired of the details (though I have tried not to encumber the text with too many of them). If I had discovered a lost economy once conducted on another planet I would not have studied it with more energy.

My enthusiasm was not enough to get me through, however. I am grateful for the financial support of the Huntington Library and Yale University for two Mellon Foundation fellowships that made it possible for me to investigate California agriculture during the early stages of my dissertation research. A Whiting Fellowship in the Humanities, generously provided by the Mrs. Giles Whiting Foundation and awarded by the faculty of Yale University, sustained my project in its final year.

In California, talented and knowledgeable people helped me at every turn. Over the course of my two years in Berkeley Robert Middlekauff offered solid advice and the best companionship. Wilbur Jacobs and John Whalenbridge asked all the right questions about this project early on. Hal Barron once suggested that the book might be called (with apologies to Eugene Genovese) "The World the Growers Made." Thanks also to Donald Pisani for reading the manuscript with care and to Bob Smith. I am especially indebted to Martin Ridge for his generosity and good humor.

I have had the pleasure to know great keepers of books. At the Bancroft Library Walter Brem gave me a quiet place to stack my books and answered my most urgent inquiries. Mary Morganti suggested how I might approach the Chevron Corporation for permission to read the records of the California Spray-Chemical Company. Many thanks also to Bill Roberts, Franz Enciso, and Dave Rez. Richard Ogar helped with photographs. Thanks also to the reference staffs of the Bio-Sciences Library at UC Berkeley and the Giannini Foundation for Agricultural Economics, and a note of gratitude to Peter Blodgett at the Huntington Library in San Marino for his remarkable knowledge of its collections. At Yale I depended on the Sterling Memorial, Seeley G. Mudd, and Cross Campus libraries. I am ever indebted to the good-natured George Miles at the Beinecke Library.

My intellectual debt to the faculty and students of Yale University is beyond calculation. Howard Lamar and William Cronon helped to shape the content of the dissertation that became this book. Howard never ceased in his encouragement for the project and remained available even while serving as president of Yale University. He is a mentor and friend of remarkable generosity. Howard's open door in the Hall of Graduate Studies and his fantastic knowledge created a spirited community of students of the West. Bill is a constant inspiration, and he exemplifies all the ideals of university life: great scholarship, wise counsel, generous teaching, and the highest standards of university citizenship. No one could ever hope to have more generous mentors. John Mack Faragher joined my dissertation committee on short notice and gave me invaluable comments and warm encouragement. He has since become a great colleague. Other colleagues at Yale also contributed to this book: Jon Butler, James Scott, Charles Remington, Carol Rose, David Weiman, and especially Robin Winks. Robert Johnston read the manuscript and offered valuable advice.

I also acknowledge the help and companionship of Randy Anderson, Edward Balleisen, Ben Carton, Roland Clements, Kelly Ditmar, Kurk Dorsey, Maia Gahtan, Emily Greenwald, Kristin Hoganson, Gunther Peck, and Frank Rocca. Leif Haase read the entire manuscript and gave me soaring encouragement along with superb comments. At the Bancroft I learned from Steven Hackel, Michael Gonzolaz, David Igler, and

David Vaught. My uncle, Bruce Neuburger, shared his remarkable experience as a picker and Communist Party organizer in the Salinas Valley. Barbara Feder helped clean up my style. Robert Kushner, Hong Kong merchant and friend since the ninth grade, always kept in touch.

Monica McCormick, my editor at the University of California Press, gave the book her full confidence and with patience walked me though an unfamiliar process. My sincere thanks to her. Richard Kirkendall and Michael Black made substantial contributions to the manuscript as readers for the Press. I only hope that I have made the most of their advice. My parents, David and Linda, and my brother, Michael, helped me through the many years of my formal education and moved my work along with their constant love and attention. This book is also for them. My in-laws, Ron and Trana Labowe, also never ceased in their care for me.

Sara knows better than anyone how much I enjoyed writing this book and how much I enjoyed leaving it behind for more important things. Our time together is when we flourish.

In the Rain's Shadow

The land was advantageous. With a gentle climate, fertile soil, and cold-water rivers, land in California offered Americans the kind of fresh prospects they had come to expect after more than two centuries of agricultural settlement. Just outside Los Angeles and across San Francisco Bay lay abundant hills for grazing and sweeping plains for crops. The snow on the far mountains melted in the spring and flowed over bottomlands in one of the largest interior valleys on earth, a place of arid summers and winter rains, a place where winds from the Great Basin dried the grass on 20 million acres to a sandy brown. At a time when Americans back East believed that the continent no longer offered distant prairies and a beckoning horizon, farmers in California used their region's natural advantages to establish a new tillage, one that did not depend upon fresh prospects and "free land," one designed to supply the consumers of an urban nation.

In the fifty years following 1880 this isolated state emerged as the center of fruit production in North America, with a rural landscape that looked like nothing the United States had ever seen before. The countryside and the industry that transformed it reflected the projects of people who called themselves fruit growers. To bypass commission merchants and hasten their produce to market, growers founded cooperative marketing associations; to keep their fruit clean of insect damage they sponsored university science and called for private chemical companies; to harvest they influenced a seasonal labor market and hired itinerant workers by the tens of thousands. Visitors wrote about towns where farmers composed the business elite and counties where the rural inhabitants earned millions of dollars by selling raisins.[1] As one traveler observed, "Intensive farming on irrigated land means a breaking away from the old methods of agriculture."[2]

Industrial farming did indeed call for breakaway methods of doing things, and no practice proclaimed a departure from the past as loudly as specialized cultivation. Single-crop farming stands at the focal point of the present narrative, holding the ground between natural advantages and the institutions that growers invented to make the most of them. This prevailing crop system maximized certain aspects of the environment but lay helpless against its own bad tendencies. The cultivation of specific crops in specific places, this book argues, brought on the industrialization of farming in California and prefigured larger changes in American agriculture. Advocates embraced "modern" methods as the violence necessary to bring the countryside into a new era. Critics cursed industrialization as a waste, a will to power, a landslide of social costs, and a lie spoken to the public in the language of Efficiency. Yet reformers and growers agreed on one thing: the entire structure depended on specialization. In this practice, institutions and environments met and shaped a remarkable rural world.

The single crop can be deduced from a more fundamental premise: that the wealth of nature is not spread evenly over the earth. Certain places enjoy certain "natural advantages," and the aim of any enterprise is to make the most thorough use of them. But these so-called advantages have nothing to do with the functioning of plants and animals in ecosystems. Advantages exist only in the imagination: they are the riches that people read into soils and climates and water. In a sense, industrial agriculture became necessary simply to sustain the particular ideas growers had about the use of land. My first task is to establish exactly what the environment offered growers as fuel for their profit-seeking imaginations. I need to explore the greater region, the arrangement of its topography, and the origin of its moisture. I am looking for patterns that will reveal how and why fruit growers used the interior lands of California to found a distinctive rural industry.

Two hundred and fifty years after the conquest of Mexico, missionaries planted their grains and olives, grazed their horses and cattle, and lived in a manner familiar to the people of southern Europe.[3] In fact, California is one of the few places on the planet with seasonal temperatures and rainfall similar to those of the Mediterranean. As the geographer James Parsons has observed, the boundaries of the state just happen to outline "the only area of winter rain and summer drought in North America with rather remarkable accuracy."[4] Although the slightest variation in altitude or soil type marked drastic boundaries between crops, this much remained constant over most of the state: rain fell in the winter, and the summer stayed hot and dry. The climate made the products of California's soil distinctive in North America and provided a unifying element in an otherwise assorted geography. What Carey McWilliams said about southern California holds true for the state as a whole: "The climate is the region."[5]

Winter rain and summer drought describes the region's cycle of moisture, but it does not explain how it works. For that, the reader must know something about two aspects California's meteorology and geology: the Pacific high-pressure cell and the rain-shadow effects. The Pacific High is a semipermanent mass of warm descending air that hangs in the upper atmosphere offshore from central California. High pressure sends air down to earth, where it blows away clouds and causes clear skies and breezy days all over the world.[6] The Pacific High blocks the polar front and storms that originate in the North Pacific from reaching the central and southern coasts. Summertime weather systems headed for the shoreline south of San Francisco Bay are deflected off the Pacific High like billiard balls and instead dampen beaches and forests in Oregon and Washington. With winter, the cell mysteriously shrinks and moves south, allowing rain to fall on Santa Barbara and Los Angeles. But if instead the High refuses to budge from its summer position (and this is not unusual), the seasonal drought continues in the seemingly benign form of perpetual good weather.

The Pacific High regulates the rain, but the mountains allocate it. Storms from the ocean drop some of their moisture on the coastal plain before encountering the Coast Ranges, a series of parallel ridges that run north and south from Los Angeles to the Oregon border. As clouds move over these low mountains and into the interior valleys, they descend, warm, and evaporate into the thin aridity in a reaction called the rain shadow. The moisture has not disappeared: it has only turned to water vapor that continues to move east with the wind across the floor to the Sierra foothills. Here, as the ghost cloud finds higher elevations, it cools, recondenses, and falls on granite peaks and Ponderosa pines as rain and snow.[7] The mountains act as a retaining wall, trapping water for California on the windward side and keeping the Great Basin a sage-covered cattle range. This frozen reservoir then holds the water through the winter until the spring when rivers bring it to the valley.

Parallel ranges traversing the state create hundreds of valleys. Much of the state's agriculture came to be conducted on these grass-covered prairies in the years after the American takeover and the Gold Rush of 1849. The Salinas and Santa Clara Valleys south of the San Francisco Bay, the Napa and Sonoma Valleys north of San Francisco, and the coastal valleys of Los Angeles and Orange Counties produced specific fruits and vegetables as a result of subtle differences in soil and climate.[8] But the state's largest and most important valley is the Central Valley, a nearly flat alluvial plain between the Coast Ranges and the Sierra Nevada running for 450 miles from the Tehachapi Mountains north of Los Angeles to Lake Shasta.[9] The Great Valley's two river systems—the San Joaquin in the south and the Sacramento in the north—meet in a freshwater delta that empties into San Francisco Bay and out the Golden Gate to the Pacific Ocean.[10] This immense interior encloses wetlands and badlands, snowmelt streams and saline sloughs, a confusion of soils, and nearly 15 million acres—diverse and fertile enough to house a nation itself.

This is the geography of blue skies that made California the greatest fruit-growing region in the nation by 1930.[11] In 1910 California's fruit and nut crop covered 646,004 acres; in 1929 it covered 3,121,986 acres or 10 percent of the state's agricultural lands.[12] This closely cultivated 10 percent generated 54 percent of the crop value statewide.[13] No state, country, or region challenged California in the commercial production of apricots, prunes, plums, table and raisin grapes, lemons, figs, almonds, and walnuts. Of these crops, California contributed between 60 and 100 percent of the total production in the United States.[14] In 1927, 67 percent of the canned fruits consumed in the United States originated in California, along with 92 percent of the dried prunes and 92 percent of the grapes. In 1928 almost 70 percent of the car-lot shipments of oranges and 100 percent of the lemons came from four counties.[15] Even in fruits known to grow to commercial quality in other places, California dominated. The Sacramento Valley, along with the counties of Fresno, Madera, Tulare, and Santa Clara, contributed 35 percent of the American peach crop in 1925 (16,251,000 bushels), harvesting more peaches than New York (1,920,000 bushels), New Jersey (1,740,000 bushels), Georgia (7,304,000 bushels), Arkansas (2,200,000 bushels), and Texas (1,750,000 bushels) *combined* (14,914,000 bushels).[16] In 1929 growers cultivated 31 percent of all the land planted to fruit in the United States, representing a farm value that contributed to California's standing as the second leading agricultural state in the Union after Texas.[17]

In this way, rural California became a set of natural advantages in the national interest—a functioning part of a larger economy and an essential resource. In the chapters that follow I will move from natural advantages to how fruit growers established specialized cultivation and the consequences of that process. The growers learned to adapt crops to local environments and ended up adapting California to a continental market.

Developments in California did not unfold in isolation. The years following 1900 were a time of anxiety about the future supply of food and uncertainty about the position of farmers in the economic life of the United States. The study of natural advantages, long the occupation of small-town boosters and experimenting tillers, ceased to be of purely local interest as economists and conservationists counted up the fertile lowlands to determine their best possible use. What follows is a discussion of land and conservation in America that might seem like a digression in a story about the California fruit business, but it is not. The fruit business derives importance from this context: the rise of a conservation ethic applied to farmland. These events mark out the place California fills in a larger narrative; they pose the problem of agriculture's relationship to the national economy; and they lead us back to those ranges and valleys in the rain's shadow.

1

The Conservation
of the Countryside

The official closing of the farmers' frontier in 1890 and the realization that arable land of high quality no longer lay in new territories initiated a period of transformation in the American countryside. Last-ditch land frenzies like the ones that tore up Oklahoma between 1893 and 1905 did little to allay the dread of scarcity.[1] The following period of high commodity prices, sometimes called the "Golden Age" of agriculture, was also a time when century-old methods of cultivation began to impede an industrial economy that could no longer expand without remaking the countryside in its image.[2] Advocates of America's industrial primacy argued that farming had become, in the stinging words of economist Edwin G. Nourse, "a belated survival of seventeenth-century modes of life in the midst of twentieth-century needs and opportunities."[3]

Not all of those who surveyed agriculture in the second decade of the twentieth century saw the destitution that Nourse did, but like him most believed that future prosperity demanded the reform of both the farm and the farmer. Some asked what kind of countryside would care for its own people by providing them with abundance, education, and society. Others asked, instead, what kind of countryside would contribute to the nation's wealth, feed its urban workforce, and increase its share of international trade. And what kind of countryside would increase in productivity without new land? Questions of social reform, industrial organization, and resource conservation all met in the "rural question," a topic urgently debated in newspapers, universities, and congressional chambers for the next thirty years.

Fruit growers in California helped to answer these questions for the nation and the world beginning in the 1880s, and they helped to determine the direction of American agriculture in the twentieth century. When conservationists urged greater efficiency in the use of limited farmland, cultivators on the Pacific Slope catalogued their soils to match crops with the most favorable local conditions. When economists advocated greater investment in agriculture, fruit growers demonstrated the commercial potential of irrigation, insecticides, and wage labor. California enters this narrative as a study in the transition between extensive and intensive farming, but it emerges as a major subject in its own right—the ultimate example of rural reform according to the new agricultural economics.

The importance of California, in other words, emerges from a story before the story that connects industrial agriculture to potent ideas and contending visions. Wrapped up in the regional narrative is a larger conflict over differing conceptions of rural reform and the purpose of agriculture itself.

The Rural Question

Agriculture in the first two decades of the twentieth century revealed the fears and contradictions of a nation wavering between a rural and an urban identity. As the number of city people approached that of people living on farms, the countryside became the subject of worried predictions, sober studies—a crisis in the minds of people who did not farm. In reality, few people in American society understood what, if anything, was wrong with agriculture; indeed, statistics from the time fail to tell a single story. Viewed one way, the numbers depict a farm economy in steady decline. The population of the United States increased from 76.1 million in 1900 to 105.7 million in 1920, a 39 percent increase. But farm output increased only 17 percent in the same period, and its growth from year to year came at a decreasing rate. Food prices in the cities hit new highs, a sign, some believed, of the sharpening disparity between stagnant production and a swelling population. Wheat jumped from 69.3 cents per bushel in 1903 to 92.6 cents the next year, and after dropping somewhat in 1907 and 1908, shot up to 99.1 cents in 1909.[4] Readers of the national census discovered that the total acreage planted to wheat decreased 15.8 percent between 1899 and 1909 but that its value increased 77.8 percent during the same period.[5]

Yet, in spite of much apprehension to the contrary, production grew by 25 percent in the first quarter of the twentieth century.[6] The amount of cropland per person did fall with the increase in population, but output per person increased until 1921, in part because the automobile came into general use, freeing millions of acres from raising feed for horses and mules.[7] Output per person averaged 40 percent higher in the years from 1914 to 1922 than it did in 1900. The number of farms and the rural population also refused to indicate a clear downward trend. Farms increased by 10 percent and country folks gained 11 percent between 1900 and 1910.[8] Still, considered as a share of the total national increase (21 percent) the rural population lagged in 1910, while the cities posted a ten-year growth rate of 34.8 percent. Rural growth nearly ceased during the next decade. Between 1910 and 1920 the number of farms increased by only 1.4 percent and the rural population by 3.2 percent. In the meantime, the cities added nearly 30 percent more people.

The appearance of disorder in the places where food came from unsettled city people. In a nation that had somehow managed to consume its "inexhaustible" frontier, no assumption about the countryside seemed safe. Higher-than-usual prices coincident with some farmers leaving the countryside spread the fear of shortages. Young men and women from farming families moved to the cities seeking better wages, setting off a period of rural depopulation that threatened to deprive farming districts of fresh leadership. Widespread soil erosion and lost fertility seemed to anticipate declining acreage and falling yields. Finally, a rise in the urban cost of living signified to consumers that the cities had come to depend on a backward hinterland that might soon languish and whither to the point that it could no longer support the progress of the nation or even feed itself.[9]

In fact wheat at a dollar per bushel meant that farmers on the Great Plains could pay their mortgages, buy factory-made goods such as shoes and tools, and keep their land in good order. But high prices provoked entirely different conclusions from the residents of the urban Northeast and Chicago. City people assumed that the higher price of food pointed to some inadequacy on the part of rural people. Complained one writer in the *New York Times*, "Farmer Brown still crops his land in a wasteful fashion, gathering one-half the harvest that moderately intelligent management would insure, while his son John has left the homestead for life in the city."[10] Another editorialist vilified farmers, calling them

relentless for gain: "A pretty good case might be made out for the assertion that the farmers are our real oppressors, the true enemies of the
consumers of their products."[11]

Distressed forecasts and the figment of scarcity reveal the capricious-
ness of public opinion following slight declines in the food supply.[12] A
drop in output in 1907 set off an urgent debate in which economists predicted short harvests all over the world and a global industrial crisis in
1908.[13] A public increasingly isolated from agriculture found it difficult to
comprehend the cadence of the seasons and the fluctuations of weather,
costs, and demand endemic to farming. With their unstudied complaints
and accusations, city people demanded that the production of hogs and
corn become as predictable as the flow of goods from factories, and that
agriculture look beyond local prosperity to serve the needs of the greater
economy. Even though the food supply was safe and growing, calls came
from all sectors for the immediate reformation of farming in the name
of national prosperity.

Yet many rural people struggled in poverty. The most respected students of agriculture agreed that the city had received the full attention
of sociologists and reformers in recent years, which had left the countryside "largely incidental and secondary" to national life.[14] Those concerned with social welfare turned an anxious eye toward rural institutions and practices and, with the fear of what they saw, fueled a movement
of reform. Wrote Theodore Roosevelt's Commission on Country Life:
"The work before us is nothing more or less than the gradual rebuild-
ing of a new agriculture and new rural life."[15] At stake was more than a
landscape of memory and the demise of the farmer as an American icon.
When leading lights in government, education, and business beheld the
miserable deprivation and stunting isolation that some farmers endured,
they saw the decline of American commercial power, malnutrition, and
political instability. As rural sociologist Macy Campbell put it, "Ulti-
mately we all go up with the farmer or we all go down with him."[16] If
the cities were allowed to drain the natural and human resources from
the land, he continued, "the foundation of society crumbles, and civilization is ripe for decay."[17]

The question was not whether to encourage change in the countryside
but how. In their desire to better the circumstances of rural people the
loosely federated Country-Life Movement found unity of purpose. Most
agreed with Roosevelt when he said that the integration of farmers into
the life of the nation was as important as the increasing quantity of their
crops: "agriculture is not the whole of country life," he said.[18] Rather

than a crisis in production and economy, the rural problem stemmed from the degradation of key institutions: the church, the school, and the farm home.[19] These became the commission's principle concern and the subject of their report to Congress. But another look discloses a more varied landscape of opinion on the condition of the countryside. Two speakers in this debate deserve attention. The views of Liberty Hyde Bailey and Edwin Griswold Nourse provide a context for changes in California during the twentieth century. Bailey and Nourse addressed a public newly interested in matters of agriculture, and they offered two opposing conceptions of the progressive countryside.

Liberty Hyde Bailey (1858–1954), the most accomplished student of agriculture of his time, dedicated his life to rural people and places. Writing in 1905, the professor of horticulture felt moved by the events of his public career to declare: "I stand, then, for the open country, for its affairs, for the trees that grow there, for the heaven above, for its men, for its women, for its institutions. . . . They are my people; with them I was born; and their problems are my problems; for them I mean to labor as long as I have strength and life."[20] Bailey was born in South Haven, Michigan, to a family that kept field crops and orchards. He graduated from the Michigan Agricultural College, worked under Asa Gray at Harvard, and returned to teach at Michigan. He accepted a professorship at Cornell University in 1888. In the 1890s Bailey published handbooks and experiment-station bulletins on every aspect of agriculture, but by the end of the decade he had become troubled by the poor conditions on neighboring farms and the disheartened feeling in many districts.

As concern over the well-being of agriculture crested, Bailey realized that the careful studies he and his students had conducted at Ithaca failed to address problems more profound than hog cholera. He saw firsthand what rural depopulation looked like: "The buildings are shabby; the grounds are bare; the fences are down; the yards are foul with weeds and litter; the cattle stand in mud; the land is hard run; the roads are poor; the inside of the house is austere and comfortless."[21] In the twenty-seven years after 1880, New York counted twelve hundred abandoned farms—all of them with soil judged to be fertile.[22] Beginning in 1905, Bailey turned his attention to the larger countryside, to the relationship between people and the land.[23]

Like the era of reform he helped to shape, Bailey is difficult to define.[24] As the director of an Agricultural Experiment Station he instructed farm families in diverse matters, like grafting apple branches and diagnosing diseased cattle. His work aimed at the improvement of farming for the betterment of the people who made their living that way. He did not, however, mistake machinery for progress, nor did he consider science the only or even the best way to improve country life. When industrialists talked about the advancement of farming they referred to machines that saved labor and organizations that would help tillers reach distant markets.

Yet Bailey was wary of everything fast and mechanical, fearing that inventions like the mowing machine, by making work easier, abbreviated the quality of human contact with nature that made farming different from city work: "My chief indictment against the mowing-machine was the fact that haying-time was no longer an epoch in the year. In the former time haying lasted three weeks; men came to the house to stay and they had stories to tell and experiences to relate, and they had new ways of doing things; now the wheel-rake had come with the mowing-machine and haying was only a mechanical labor and it was over in less than a week."[25] This could easily be taken as a song of lost innocence, but it is more. Though Bailey did not recoil from technical solutions (he was a scientist), he believed that any modification in the methods of agriculture should find some compromise with the deliberate pace of the countryside, a landscape he considered essential to national survival for reasons other than its capacity to produce cheap food. Improved tillage quickened the cadence of rural life and expanded the scale of farming. Caught between his desire to ease the strain on farmers by helping them provide more for their families and the impulse to adopt new technology that was the creed of his times, he tried to find some middle ground where the tools of progress might work side by side with the farmer as he smelled newly churned soil and felt the soft seed heads in his wheat field. Technology should help farmers do what they have always done, he believed—not change their lives.

Bailey worried about the effects of the mechanized farm on the outlook of the rural population: "These changes have been of two diverse kinds," he wrote, "those affecting farm practice, and those altering the sentiment of the farming people."[26] The second was the more disturbing, for nothing troubled him more than the possibility that country people themselves might be transformed. New machines might lead to a modest rise in income, which would entice farm families to acquire expensive furnishings for the home. The women would want the newest clothing and the kids would become bored with milk cows and clover. An urban lifestyle demanded money beyond what the soil could reasonably provide, for "the productivity of the land is capable of only limited increase."[27] Bailey saw farmers straining to maintain the standard of living of those who did not farm, suffering the sting of inferiority and changing farming in order to satisfy their wants. "Even the farming itself is changing radically in character. It ceases to be an occupation to gain sustenance and becomes a business. We apply to it the general attitudes of commerce."[28]

Although Bailey recognized certain discoveries as advances in the science of tillage, he also saw them as denigrating to the almost mystical compromise between humans and nature. Machines might do more than spoil peace and quiet. They might cause rural people to lose sympathy with the ecosystems that provided for them. None, perhaps, but John Muir and Aldo Leopold ever articulated a more integrated relationship between humans and nature, nor one more demoting to the importance of the former: "We are parts in a living sensitive creation. The living creation is not exclusively man-centered: it is bio-centric. We perceive the essential continuity in nature, arising from within rather than from without, the forms of life proceeding upwardly and onwardly in something very like a mighty plan of sequence, man being one part in the process. . . . We can claim no gross superiority and no isolated self-importance."[29] He called the farmer a naturalist, a steward of creation, and believed that to remove the tiller's commonplace respect for providence would render him reckless for gain.

With Bailey's views on the natural environment we can close a circle in his thought: rural institutions (church, school, and home) would save the countryside from industrialism because these pillars of society held farmers and their humble, conservationist values on the land. And just as Bailey perceived the continuity of nature as "arising from within rather than from without," so too did he see the continuity of the countryside—its improvement, its reform—as arising from within. The country, Bailey wrote in 1903, "exists for itself and maintains itself."[30] Conserving it did not mean mowing machines and industrial revolution. For Bailey it meant doing what was necessary to keep the rural population self-sustaining. The children had to learn to love nature, and their parents had to learn how to make the farm pay in the form of material comfort and in the less tangible currency of a Good Life.

These are some of the ideas that Bailey brought to his position as chair of Roosevelt's Commission on Country Life. The commission's report, delivered to Congress in 1909, expressed much of his philosophy. It alluded to industrial farming and California agriculture by criticizing "undiversified one-crop farming" and farmers who raised none of their home supplies. It called on cooperative marketing associations to perform a social role, saying that many acted "with no thought of the good of the community at large" but simply for their stockholders in the manner of commercial corporations.[31] In this and other writings Bailey criticized few specific tools or technologies, and he provided few answers to farmers who felt compelled by difficult circumstances to adopt new methods. He insisted that industry and agriculture could never mix, calling the highly commercialized cultivation made popular in places like California "freak farming" and bound to fail, but he declined to speak to farmers with words they could put to work.

In the years following the publication of the report, thinkers with different ends in mind entered the debate. They tended to see agriculture as a discrete function of the economy, detachable from rural life. Among them was Edwin Nourse.

Utterly unromantic about agriculture, Edwin G. Nourse (1883–1975) did not equivocate or qualify when it came to its reform. Nourse dedicated his long career to an industrial revolution in the American countryside. Born in Lockport, New York, he graduated from Cornell University in 1906. A Ph.D. from the University of Chicago followed in 1915.[32] Nourse accused farmers and economists of failing to understand the significance of agriculture in the complex society fast upon them. Farmers, he believed, needed to cast off the assumptions of the "free-land period"—the nation's extended adolescence when agriculture lived in ignorance of economic forces. The frontier, Nourse wrote in 1916, "tended both to direct men's minds away from purely economic theorizing, and to mislead them when they did attempt to pass strictly economic judgments."[33] The following period of rural depopulation that induced so much alarm among sociologists and politicians was, to Nourse, a necessary "bleeding" of the patient. Farming in the new era required stronger stuff than frontier yahoos.

In Nourse's view farming had no other purpose than to serve the industrial expansion of the United States. "Agriculture," he wrote, "can rightly claim no . . . special position in our economic system."[34] Unrestrained by the "churlish agrarianism" he disdained, Nourse advanced the argument that "American agriculture must be thoroughly reorganized upon the basis of modern industrial efficiency." He brought the principles of business enterprise to the study of agriculture and called for "a change in personnel and organization, in fact, a thorough recasting of the whole business of farming." Nourse imagined a tiller who would be "heir to all the complex knowledge that the world holds," a farmer who could "fill the very large place marked out for him in our high-pressure industrialized regime." [35] To remove pastoral sentimentality from the farmer would render him a manager of food production, an engineer whose materials consisted of soil, climate, and moisture. Nourse did not want farmers who insisted on a definitive "rural life" but those who would lend their land and capital to the industrialization of agriculture.

Nourse took farming as a weapon in the quest for American preeminence. He argued that the end of World War I initiated a new era of international competition that the United States must be poised to win. Any sector of the economy acting as a drag upon the ship of industry should be hauled in or cut loose: "All those interests which in any way run athwart that line of development must impose a self-denying ordinance upon themselves or be put by a strong hand back into their humble place of servitude." [36] Manufacturers "must strip themselves of all hampering influences" in order to win the new game of economic imperialism. Nourse had equally harsh words for the trust-busting government whose "foolish zeal to regulate business organizations" threatened these organizations' ability "to adopt whatever commercial practice [that] may conduce to their success in the face of foreign competition." This was war under another name, and just as war required sacrifice, so too did the commercial struggle require that the comfort and even the survival of some should be given over to expedience. As Nourse put it, "Neither the maintenance of economic standards at home nor a living wage and decent treatment for sailors afloat must be allowed to handicap these knights of trade so unselfishly eager to set our flag over every commercial rampart of the world." [37] This is the final context for farms and farmers in the world that Nourse imagined: subsumed under the flag of a corporatized world, enlisted into the service of muscular commerce.

Bailey and Nourse illustrate a deeper tug-of-war between the city and the countryside for cultural dominance. Since Benjamin Franklin and Thomas Jefferson offered two distinct landscapes where American citizenship could be practiced as public virtue, Americans have considered the implications of a nation based on the city or the country. Leo Marx has well stated the American dilemma of preserving some essence of the country while fast invading it with commercial thinking. "The objective, in theory at least," Marx writes, "was a society of the middle landscape, a rural nation exhibiting a happy balance of art and nature. But no one, not even Jefferson, had been able to identify the point of arrest, the critical moment when the tilt might be expected and progress cease to be progress." [38] In time, Marx continued, talk of a balance between the two spheres lapsed into vacuous rhetoric and ceased to command attention. After the Populist campaigns fell apart in the 1890s no organized movement arose to challenge the rush of steel and smoke—and all it represented—from rolling over the countryside.[39] This is the way things stood when Bailey and Nourse and hundreds of others considered the future of the middle landscape in the second decade of the twentieth century. Both felt the gravity of the changes they weighed. Factories, mills turning on rivers, railroads crossing the prairie, banks, grain elevators, and boards of trade all seemed to comprise a prologue to the coming of industrial order to rural production itself.

Neither Bailey nor anyone else seemed capable of slowing the pace of change. Nourse believed that the countryside had no value apart from its capacity to produce cheap food, a view that justified any innovation that offered to increase output. For his part, although Bailey urged Americans to find the point of arrest—when progress would cease to be progress—all his eloquence came down to a plea that farming might change and still stay the same. Bailey's position criticized but made no argument to block the advance of machines and corporations and commercial relations. He understood that the desire to use technology—not the machines themselves—drove industrialism.[40] By the 1910s Americans searching for an example of modern farming could look to the orchards and vineyards of California. Farmers there pursued efficiency and profit and never stopped believing in progress.

Still, we are not yet ready for California. Questions of reform existed quite apart from the immediate work of farmers and the economic forces they faced.[41] Of special importance to farmers was how best to use the limited resources of land—its climate, soil, and water. There was nothing tributary about land utilization to the formation of the industrialized countryside in California; indeed, nothing was as critical to its development.

Why Regions Specialize

In order to survive as participants in the national economy farmers needed to reconsider how they used their fundamental raw material: land. Beneath talk of new methods lay the land itself and plenty of anxiety over its capacity to produce. Agriculture after the frontier presented a problem as old as economics. The population of the United

States doubled from 50 to 100 million between 1880 and 1920, and the 200 million people predicted by 1960 seemed to foretell shortages.[42] In response to the population trend and the dwindling supply of quality soils, conservationists and economists began to advocate greater efficiency in the use of farmland. Only by making the best use of the best land by planting the most productive crops would farmers and city people prosper together. Farmers in the various regions of North America began to specialize in the plants they could produce in high quantities by virtue of the particular natural advantages outside their doors.

Land utilization took on special urgency after 1900. American farmers struggled to make a living under old ideas, and they were on the move. Seeking to cash in on high commodity prices in the years preceding World War I, many farmers went to the Great Plains in search of fresh land. More new homesteads appeared in the West between 1900 to 1920 than in the last twenty years of the nineteenth century.[43] Westering tillers pursued a single prize: larger surpluses, or more of the produce left over at the end of the harvest after costs and subsistence. During the period from 1820 to 1870 American farmers made a quiet transition when they began to send a larger portion of their crops to city markets than they consumed at home. As finished goods like clothing, tools, and sugar became accessible and indispensable to rural families, surplus took on greater importance until it became the object of farming itself.[44] The drive into new territories like the Great Plains in search of surplus caused the kind of expansion that conservationists most feared, and it explains the anxiety behind land utilization.

Generations of American farmers believed that harvesting larger crops meant moving to richer soil. They arrived at this conclusion, in part, because of the poor position they occupied whenever they went to market. One of the most important things to know about agriculture as a *household* economy is that farmers almost never set the prices for their commodities. No matter what it cost them to hoe and plow, no matter what they paid in high freight rates or poor weather, they could not make up the loss by raising prices. When they arrived at country stores in 1830 or grain elevators in 1910, farmers always met many more sellers than buyers. Since one bushel differed little from the next, all the sellers in the same place at the same time came away with close to the same price.[45] Their only recourse in good times and in bad was either to lower the cost of production or make more surplus. Making more surplus, in turn, required farmers to cultivate more land. Fresh acres in places like Ohio and New York could be difficult to find, and many farmers did not know how to cut their costs to reach a point of profitable return. So they suffered low incomes and saw their children move to Chicago. Those with the gumption to pick up and move looked to the unbroken soil of the Great Plains as an opportunity they could not resist.

This is the assumption behind *extensive* farming, the familiar process of searching out greener acres when fertility failed or when falling prices pressured a farmer to enlarge the homestead. Extensive expansion proceeded through the application of old methods of tillage on new land, and it brought settlers to every corner of North America.[46] But by the early twentieth century farmers could identify only one more blank spot on the map where they could still stake out virgin land. When international demand for grain during the Great War sent prices into the blue sky, opportunistic landsmen came to the Great Plains ready with their steel plows and frontier expectations. Settlers came for cash, understanding little about the stark and windy world they presumed would provide for them.

Their arrival could not have been less auspicious. The first wave of settlement began in the 1870s and retreated within about twenty years of its peak. By 1900 scattered counties in the southern plains had lost 30, 60, and even 90 percent of their sod-house populations.[47] One researcher into methods of dryland farming concluded about the Great Plains that the settler on a section of land (640 acres) stood "a fair chance" of surviving, while the settler on a quarter section (160 acres) could expect "almost certain failure."[48] John Wesley Powell, a government geologist and ethnographer, was one of the few to understand the region for what it was: marginal land. His 1878 *Report on the Lands of the Arid Regions of the United States* asserted that without irrigation no certain settlement could be established there.[49] "Periods in which settlement was rapid, energetic, and general," wrote a government geographer, "have alternated with periods when abandonment, desertion, and return were almost as rapid."[50]

Then it started all over again. Between 1900 and 1920 the northern plains felt a 300 percent rise in population. Improved land went from 1 million acres to 27.3 million acres in the same two decades, including 6.6 million acres of wheat.[51] But the new burst of settlement did not change facts. The Great Plains saw the twilight of extensive farming, the last attempt to extend a kind of tillage that no longer made sense in a world without "free land." Extensive expansion and the worn-out soil it left behind convinced students of agricultural economics that American farmers proceeded under false assumptions.

The rural question could also be phrased in the language of conservation. Public concern for resources like forests and rivers also extended to farmland. What was cruel wilderness to the generation that fought at Lexington and Concord became valued resource to those who fought at the Marne. In conservationist literature like Rudolf Cronau's *Our Wasteful Nation* (1908), Americans learned that their natural wealth, far from being as limitless as the horizon, could in fact be consumed into oblivion.[52] Conservation transformed nature into capital (indeed, into a natural inheritance) that earned an interest income that each generation could responsibly spend on itself. The idea was not to cease lumbering and mining, but to manage these activities to conserve the resource. Now the harvesting of forests and the mining of metals—once called progress—became "extractive," and the cultivation of unbroken land as a means of expanding production—once called "improvement"—increasingly appeared wasteful and haphazard. As James J. Hill told farmers in 1908, the time had come for deliberate agriculture and careful cultivation: "The man no longer deserves the name of farmer who conceives of his industry as a scratching of the earth, a hit-or-miss sscattering of seed and a harvesting of such yield as soil and weather may permit. That is not farming, but a game of chance."[53]

In the twentieth century, if their soil gave way to erosion and exhaustion, farmers had few places left to go. The only fertile land lay either at the bottom of swamps in need of draining or on desert plains with insufficient rainfall. The first required too much capital for the typical farmer to consider; the second required a suspension of common sense. Since the founding of James Town the expansion of agriculture had depended on the cultivation of new land. How would this expansion continue in the age of limits? Farmers needed to grow more food on land they already owned. Hereafter, agriculture would advance *intensively*, or through the application of capital, technology, and new methods to existing farmland in order to attain the fullest possible use of scarce resources.[54] The frontier, in short, had to be redefined as the potential productivity of already-cultivated land, and the tools of this new settlement also had to change. Capital in the form of irrigation, fertilizer, insecticide, labor-saving machinery, and more productive plants in effect "added land" to farms by increasing yields without geographical spread.[55] Limited resources forced farmers to look away from the horizon and "refinance" the fertility of their soil.

This is the idea that transformed the California countryside and made it the leading example of a new kind of agriculture. Yet the first rule of intensive cultivation had nothing to do with fancy inputs. As one lead-ing California grower expressed it, "The most profitable production of a soil and climate is that which is most aided by the natural forces."[56] The natural forces consist of the many qualities of *land*, an omnibus term that includes soil fertility as well as mineral composition, location, topography, moisture, and altitude. Land poor in essential elements could not be expected to yield profitably no matter how much capital and labor might be applied to it. Choosing it carefully and planting it correctly meant the difference between a comfortable living and years of hard toil.

Precise planting exploited the virtues of land. The search for compat-ible plants constituted a kind of running negotiation between tillers and the earth. This process resulted in regional distinctions like the corn belt, the hay and dairy belt, and the cotton South.[57] These regions took shape during periods of settlement and experimentation when farmers selected species most likely to match local conditions. New after 1900 was the ur-gency of the discussion and the significance of the outcome. Productive and abundantly watered farmland stood in limited supply, and just as trees and rivers became precious gems in America's natural treasure chest, so too did farmland come to be classified and inventoried to ensure its maximum use. There would be no more time for the deliberate experi-ments of the settler. Now farmers had to plant well in order to feed the people who did not farm.

Maximum use became a conservationist catchword, an academic movement, and a source of public policy that aimed to determine the most advantageous physical conditions for all major crops. Its goal: to match plant and environment with the greatest possible accuracy. It re-ferred to any attempt at growing the right crop in the right region to realize the highest possible yield for the least amount of capital and la-bor. Writing in the teens and twenties, a group of economists and geo-graphers held that the end of the frontier marked the beginning of a new era in which farmers would have to pay close attention to subtle-ties of climate and soil and cultivate in strict accordance with them.[58]

First among the students of land utilization was Oliver E. Baker. A geographer and economist who worked under Lewis Cecil Gray at the Bureau of Agricultural Economics, Baker explained the economic and environmental basis of regional specialization in California and North America during the 1910s and 1920s.[59] Building on the work of David Ricardo and Johann Heinrich Von Thünen, he argued that in a world linked by railroad and telegraph, the physical differences between regions became increasingly important in determining their use. Regions once isolated from each other now came into direct competition for dominance over specific crops. Only farmers in regions with the most advantageous natural attributes could expect to win this competition, since they harvested more and better produce at a lower cost and were able to afford more intensive methods than farmers in less fortunate places.[60] For those poor farmers growing the wrong crop in the wrong climate, Baker predicted resource inefficiency and financial ruin.

Baker and Gray realized that no matter how much marginal land went into production, it would never result in a stable and productive rural economy. Dry, insect-ridden, and disaster-prone land made for insolvency and privation no matter what kind of fancy machines and chemicals a farmer employed on it. No input superseded the quality of the land. If the cost of inputs exceeded the value of the additional yield they brought, the purpose of intensive agriculture was defeated.[61] Baker's own tour of the blighted countryside west of the hundredth meridian confirmed his supposition that the future of American agriculture depended on informed land utilization: "One who has seen the struggles and failures of the young, ambitious, and often capable farmers in the drier portions of the Great Plains becomes convinced that the good land is occupied and that land both cheap and fertile no longer exists in the United States."[62]

The only solution to the dilemma of an increasing population and the depletion of fresh territory, Baker believed, was the combination of intensive methods on the best soil available. Conservationists and economists argued that productive agriculture needed farmland capable of sustaining high yields, farmland endowed with high fertility and mild climate, farmland worthy of the capital that farmers in the twentieth century would invest in it. Baker's observation of the Great Plains suggested to him that American farmers rushed unthinking to waste their energy and capital on marginal land, and he forewarned: "The waves of population are beating against the barriers of adverse physical conditions all along the shore-line of settlement."[63]

Baker did more than write dark prognostication, though in order to appreciate the more subtle aspects of his thinking we need to seek out one of his most important sources: David Ricardo, the eighteenth-century economist who offered an influential theory of rent.[64] Ricardo's conception of land and capital provided the underlying theory behind California's astonishing rise. Simply put, he proposed that in a world of sharp competition, only the people on the best land prospered.

Ricardo began with the premise that no two pieces of land have the same qualities. Excellent soil yields a bumper crop under the care of a novice, while poor soil fails to pay when cultivated by an expert. In Ricardo's conception, rent is the difference in yield between two unequal pieces of land after an equal amount of labor and capital is invested in each. Imagine a world with three kinds of land: best, average, and poor. Where settlers cultivate only the best land rent does not exist. As long as each owner invests the same amount of labor and capital, each receives the same yield at harvest time. Because rent is a comparison of the inherent productive power between two or more tracts, it only exists where some land is of inferior quality.

When more people arrive they find all the best land taken up. These late comers are forced to cultivate average land, and differences immediately become apparent. Where the best yields one hundred bushels per acre, the average yields only eighty. The twenty-bushel difference, according to Ricardo, constitutes "rent" on average land. Note: it makes no difference that owners of average acres do not literally pay a landlord. A rent of twenty bushels represents the "cost" of owning average land as well as the "income" derived from owning the best land—it need not be expressed by an actual exchange of money. When the country finally fills up, desperate farmers try to make a living on the poor land, but their sweat earns them only sixty bushels per acre while the same resources bring one hundred bushels to the fortunate owners of the best land. In other words, think of rent as a measure of the relative productivity of land—its level of output for every dollar of input.[65] Farmers on marginal soil spent more to harvest less, and the countryside suffered.

Oliver Baker applied Ricardo to North America. With railroads and telegraphs sewing up the continent, long-isolated regions became as close together as two parcels in Ricardo's English countryside, and their productivity could be compared. The only farmers who could hope to prosper in the fast and unforgiving continental market—the only farmers who could afford transportation and all the uncertainties of agriculture and long-distance trade—were those who achieved the greatest possible productive efficiency on the best land available. Farmers with high-quality soil and mild climate stood a better chance of finding profit even at low prices. The theory of land utilization held out the possibility that farmers far from central markets could compete with farmers just outside the city center.[66] This is the principle that made California the greatest truck garden in North America in spite of its isolation from Chicago and New York. Fruit growers made the most of excellent conditions and used their profits to reach consumers far beyond the Sierra.

One more element is still missing from the mix that made the California countryside: specialization.

Land marginal for one purpose might be perfect for another. Farmers across the continent could safeguard their prosperity by identifying the plants best adapted to their regions.[67] If tillers everywhere planted according to the unique qualities of their local environments, then all might find themselves owning the "best" land for specific crops. For example, Massachusetts, having long since lost its advantage in wheat growing, discovered that cranberries lying in bogs could be surprisingly profitable, and few other states could grow them. Apples came from hundreds of locations throughout the continent, but commercial production concentrated in a few places like central Washington and the lake counties of New York. Cheese production in Wisconsin shifted to those regions with fewer than 150 days in the growing season, because cool and short summers discouraged bacterial growth and made for better pasturage. The sandy and infertile soils of the Upper Lakes Region, an area of cutover forests and abandoned farms in the 1920s, reverted back to forest. And farmers in California realized that many of the fruits they planted in alluvial soils grew to fantastic quality and produced large crops. Products like lemons and raisins could be attempted almost nowhere else. Cultivation close to regional natural advantages promoted the most efficient use of land, effort, and capital, resulting in larger harvests, lower costs, more food, and more affluent farmers.

Yet what makes this idea so important to the story of California is that it never worked this easily. Another economist explained why advantage always confronted disadvantage in the creation of a vast agricultural economy. Johann Heinrich Von Thünen looked at the world with a different set of questions than David Ricardo and came to different conclusions about the prospects of agriculture far from the point of consumption. Whatever their natural advantages, said Von Thünen, regions isolated from markets faced the prohibitive disadvantage of distance. He imagined a fictitious state with a vast agricultural hinterland surrounding a single town. He broke up the landscape into a series of production zones designated by concentric rings emanating from the town. The farther away one settled, the more expensive it became to bring the crop for sale and the lower the profit. The outermost ring marked the area of subsistence farming where the cost of taking grain to market summed to more than the value of the produce itself. The innermost ring indicated the region of intensive fruit and vegetable production, a space close to the center beyond which "delicate horticultural products . . . would not survive long journeys by wagon."[68] Natural advantages meant little in a world where distance dominated agriculture. Distance destroys value, Von Thünen argued, so it determines where commodities are produced.

Von Thünen did not count on the railroad, an invention that eased the misfortune of distance and transformed the American landscape into everyone's backyard.[69] Distance still mattered, but rather than a wall in the path of commerce it became a significant but surmountable obstacle. Farmers who produced a commodity of sufficient value to assimilate the cost of sending it could sell their crops in faraway places. Still, the railroad did not convince everyone who considered the problem that agriculture could be conducted without regard to location. George F. Warren, a colleague of Liberty Hyde Bailey's at Cornell University, regarded production near to consumers as one of the basic laws of farming.[70] Warren cautioned his readers that "next to soil and climatic limitations the freight and express rates and cost of handling produce are the most important factors in determining the type of farming [in any location]."[71]

Oliver Baker also acknowledged that the cost of shipping threatened the benefits of regional specialization. The trend over the previous decades had been toward specialized regions "where geographic conditions were most favorable, more or less *without regard to State or even national boundaries.*"[72] But should railroad rates become too expensive, "the effect would be to . . . discourage the development of highly specialized districts of production remote from markets."[73] In fact, transportation composed only one among many other costs that threatened the viability of specialized agriculture. *Anything* that raised expenses or in any way diminished the quality of produce—whether difficulties in marketing, spoilage in transport, lack of plentiful labor, or infestations of insects—could erase the profits earned by superior natural advantages.

Specialization offered a strategy for prosperity to regions across the continent, but it was not without risk. The benefits of place lived in unceasing conflict with the perils of high costs. If a balance could be struck between them, city people would eat from the productions of farmers throughout North America. Agriculture would then join in the technical sophistication and regional interdependence that increasingly characterized the industrial economy. A well-fed nation with affluent farmers depended on the forces of Ricardo to beat down Von Thünen's determinism.

This is the context for the rise of California. How fruit growers in the Far West played the balance between advantages and costs tells a story about agriculture in the twentieth century, a story that maps out the industrial countryside.

Relative Advantages:
California's Wheat Bonanza

No region illustrates the fortunes and reversals of special-ized agriculture in the twentieth century better than California. To a greater degree than almost any other state, California turned its agricul-ture into a model of intensity. Fruit growers from Sacramento to Fresno to Ventura became some of the wealthiest farmers in the world and cre-ated a countryside astonishing in its productivity and disturbing in its implications.

Thirty years before Oliver Baker published his essays on land utiliza-tion, boosters on the Pacific Coast considered how to use their state's soil and climate to master commerce in the wider world. Speaking before the State Agricultural Society in 1889, William H. Mills, a land agent and offi-cial for the Southern Pacific Railroad, described the interregional com-petition that would remake rural industry in California. In markets all over the world, Mills said, "the competition of soils and climates is im-mediately present. . . . In these markets, we see the fertility of soils and the favoring conditions of climate competing with the environment of every other portion of the world. . . . In every market there are imme-diately present the effects of the systems of labor, the methods of pro-duction, the favoring conditions of soil and climate; they meet face to face; distance no longer divides them. Their economic presence has be-come the equivalent of physical contiguity." [74] Mills announced that Cali-fornia had already won this face-to-face competition. One need only con-sider the state's exemplary agricultural conditions and its products of the highest quality, "and you will perceive that the orchards of California and the vineyards of this State are in immediate competition in the mar-kets of the world . . . ; that it is possible, therefore, for us to become the orchard for the whole world."

As for the Von Thünen-like perils of a far-flung trade, Mills claimed that California had made a fool out of distance. A recent reduction in the price of freight from one thousand dollars per carload to two hundred dollars, coast to coast, inspired him to argue that the Southern Pacific Railroad had transfigured the earth to the benefit of California's rural in-dustry. Because it now cost as much to transport goods from San Fran-cisco to New York as it did to ship from the Missouri River to New York, Mills concluded that California lay only as far away from the greatest market in North America as the Missouri River. By this same calculus he asserted that the new rate policy had effectively "removed Chicago to within four hundred and fifty-five miles of the gardens of California."[75] Technology laid waste to space and "placed orchards, physically distant from each other, side by side in a commercial sense." Mills dreamed the big dream of California's farmers and fruit growers: that a beneficent nature and a little largesse from the Southern Pacific could bring the garden produce of the Far West to the great centers of consumption.

Distance did not fold into time, and California agriculture did not rise in a straight line of progress. In fact, the state's rural economy underwent a transition in the 1880s, one that revealed the sensitivity of natural advantages. Fruit became California's second multimillion dollar harvest, not its first. That distinction belonged to wheat, the most versatile of all old-world grains. The same shift from extensive to intensive agriculture that Oliver Baker urged in the 1910s had been under way since the 1880s in California as fruit orchards on small irrigated tracts replaced wheat fields the size of counties. The wheat "bonanza" illustrates both the wealth that derived from ideal agricultural conditions and how relative to time, place, and circumstance those advantages could be. For what is significant about the wheat boom of the years between 1870 and 1900 is not only its success but its demise.

Wheat became important in California for the same reasons it did elsewhere. It is hardly a discerning plant. It can be grown with very little care, in many different soils, in almost any climate with about twenty inches of rainfall. Wheat can even be grown in snow. These characteristics made it particularly well suited for places with slow and uncertain communication with the wider world. In addition to its role as a staple for the subsistence farmer, wheat could be traded with confidence. It kept well over many days of travel, was relatively lightweight compared with corn, and could be sold anywhere people were in the habit of eating. Low-paying but hardy, it proved to be the perfect cash crop for lonely California and had already become important by the 1840s when John Sutter planted it at New Helvetia in the Sacramento Valley. Portions of the northern valley were virtually given over to wheat as farmers realized that it could be transported thousands of miles in the ballast of the same ships that brought miners and supplies to San Francisco. It was not the Sacramento Valley but the drier and more isolated San Joaquin Valley that saw the apex of wheat's career on the Pacific Coast.

Early American visitors to the San Joaquin Valley noticed nothing advantageous about its miles of dried grass and afflicting heat. William Brewer, a Yale University geologist, traversed it during the Civil War when a lasting drought killed thousands of cattle. He and the other members of the Whitney Survey described an environment seldom seen by easterners of the 1860s. It was, in Brewer's words, "a desolate waste—I should call it a *desert*. . . . The soil was barren and, this dry year, almost destitute of vegetation. . . . We stopped at a miserable hut, where there is a spring and a man keeps a few cattle." Brewer called it a treeless void with endless plains, stinking alkali sloughs, and shanty homes—an unlikely place, he thought, for any permanent settlement.[76]

An eastern geologist, however, could not be expected to appraise this land with the same instinct for possibility as a western speculator. Following Brewer by only a few years came people who looked at the same arid prairies and imagined wheat fields. Land speculators began to visit the San Joaquin after 1867, the year the Southern Pacific Railroad announced that it would build a branch line down the valley connecting San Francisco and Sacramento with Los Angeles. One of those who came to see for himself was a sheepherder named Moses Church. Church understood a few things about desert climate and observed that "where alfilaree grew so abundantly and sunflowers 10 feet high, wheat would also certainly grow." He realized that if the grain could be cultivated on the same seasonal cycle that made the grass come up green in winter and dry by June, a wheat economy would flourish.[77] Farmers could sow from December to March and harvest throughout the summer.[78] The resulting product was as hard and dry as the climate that produced it, enabling it to travel fourteen thousand miles in ship bottoms to its principal market, Liverpool, England.[79]

The natural environment provided the basis for agriculture in California, but the social environment provided its impetus. The desire for gain through speculative investment—born of the Gold Rush of the 1850s—characterized farming no less than mining.[80] In the years after 1849, when it became apparent that the gold in the Sierra Nevada would not last, merchants, bankers, and former miners purchased land from the federal government with the expectation that the future demand for farms would increase the value of their holdings. Some acquired extraordinary acreage through dubious means.[81] The owners resided in San Francisco where they became members of the city's commercial elite. They included Isaac Friedlander, a merchant and socialite known as the "Grain King" who sold more wheat to England than anyone else in the state; Henry Miller and Charles Lux, two butchers turned cattle ranchers who by the 1860s owned one hundred miles of the San Joaquin River; William C. Ralston, who presided over of the Bank of California; and Lloyd Tevis, who directed the Wells Fargo Company. William S. Chapman purchased college land script under the Morrill Act of 1862 and, by

1871, owned 650,000 acres of California and Nevada. By 1881, he owned 1 million.[82] These investors and speculators initiated the agricultural development of the San Joaquin Valley. With the Southern Pacific Railroad moving surely toward Los Angeles and with a market for wheat in England, land once judged sterile for its lack of trees and worthless for its lack of water became a great field of wheat.[83]

The result was a sight that left many visitors amazed. A typical wheat harvest on L. A. Richards's ranch began on June 1 when six header machines operated by crews of American and Portuguese workers cut 270 acres and threshed three thousand bushels in a single day.[84] The operation required forty-three men, one hundred horses, and took one month to complete. A writer for *Appleton's Journal* climbed to the roof of a machine shop in the San Joaquin one June in the 1870s and saw, "into the faint recesses of the foot-hills, and just as far to the north, and also as far to the south, . . . one unbroken sea of yellow grain."[85] He observed 32,000 acres about to be harvested, an operation he could compare to nothing but an army on campaign: "There is no suggestion of gentleness, or grace or poetry in the whole wheat field. . . . All is ingenuity, precision, order, force."[86] The crop traveled on the San Joaquin River to San Francisco, where merchants shipped it around Cape Horn to Liverpool. Isaac Friedlander, who at one time controlled 214,000 acres in California, netted close to $250,000 in a single season.[87]

But the reign of the grain kings, though remarkable, was brief. By the early 1870s it was already apparent that wheat depleted the fertility of the soil and kept the rural population small and isolated. Critics abounded. In *Our Land and Land Policy*, published in 1871, Henry George used the wheat bonanza as his principal example of the evils of land monopoly. California, he observed, with an area larger than many European nations, "does not contain the population of a third-class modern city."[88] Wheatgrowing strangled economic development by preventing widespread settlement. One booster in Fresno County put the problem in these terms: six sections (3,840 acres) planted in wheat might earn a single family $25,000 a year. The same six sections planted in fruit and divided into 20-acre lots would support two hundred families, each earning about $3,000 a year.[89] Charles Nordhoff, an early popularizer of life in California, described the bleak homes and hand-to-mouth existence of many who came to the San Joaquin—like miners to the placers—to make quick money and leave. Disturbed by the whole business he wrote: "It is not a pleasant system of agriculture, nor one which can be permanent."[90]

Although the critics of the bonanza battled against it, wheat died from its own excesses and the changing conditions of world commerce. Farmers in other places also began to send wheat to the British market, and the natural advantages they enjoyed could be felt in California. The president of the State Agricultural Society admitted in 1890 that California wheat had finally come into "direct competition with the lands devoted to like production," and that local counties now fought distant countries for the same consumers.[91] When Argentina, India, Russia, Canada, and Australia became suppliers they drove the price down. The cental (a British measure for a short hundred weight) fell below $1.50 for the first time in 1883 and never recovered.[92] A decade earlier Charles Nordhoff interviewed a farmer who realized only one and a half bushels per acre, but with 700 acres under the plow and prevailing prices in the early 1870s the man still made money.[93] As long as prices remained high, even yields this low paid off, but with the drop of the 1880s extensive cultivation felt a fast decline. Cries rose as prices fell and the distance between California and Liverpool seemed to increase.[94] Perhaps, some speculated, the Southern Pacific would reach New Orleans, allowing wheat to be shipped from the Gulf of Mexico. Perhaps a canal could be dug across the Isthmus of Panama—a potential savings of ten thousand miles on the ocean voyage.[95]

Neither would have saved the wheat business. Wheat culture wasted its resources and fell into decline in spite of new machinery. Inventions like the Stockton gangplow allowed one person to break the soil on a wide path and scatter seeds behind. Combination harvesters cut, threshed, and eventually sacked grain in hundred-pound bags.[96] Though astonishing in their power, these represented old methods of tillage refashioned for the monstrous scale of the San Joaquin Valley. Grain growers did nothing to improve the fertility of failing soil. Instead, they mined the dirt for its last dollar and brought the wheat boom to an end after less than twenty years. With no more cheap land to bust, farmers abandoned their dismal acres, leaving behind a desolation that was the agricultural equivalent of ripped-out gold hills and dug-out river bottoms. As terrific as the bonanza had seemed in the 1870s, two decades later the San Joaquin had become marginal land for the purpose of growing wheat.[97]

The end of wheat taught new lessons. California lost its advantage because it could no longer raise sufficient quantities cheaply enough to justify the expense of exporting it. Extensive agriculture had no way to enlarge its yields in response to falling prices except to replicate its old pattern of spread for temporary gain. California wheat ran its course, no longer able to compete with better situated regions. The natural advantage—always contingent on environment, competition, and methods of tillage—had been lost.

When landowners contemplated a new use for their idle lands one alternative seemed most compelling. Boosters in newborn railroad towns like Fresno and Tulare considered whether the water of the San Joaquin River might be better used to grow crops than to float them to San Francisco. Experiments with irrigation near Fresno and elsewhere convinced William Chapman, for one, that water would soon found a new rural landscape. If canals could divide the flow of rivers the land could be subdivided and sold. The San Joaquin Valley, he wrote, "is calculated, at present, for *making money* by raising wheat on a large scale." Once exploited and exhausted for this purpose, he predicted, the valley would "be subdivided into small farms and vineyards."[98] Out of a desire to dispose of land they could no longer rent for wheat or cultivate themselves, speculators and merchant-farmers became land developers, and irrigated real estate became their next venture.[99]

The flow of diverted water would bring with it a thriving rural economy. William Smythe represented irrigationists throughout the West when he wrote that water promised to turn the desert into a trans-Mississippi New England—a nation of independent farmers. Yet Smythe clearly expected that farmers in this agrarian utopia would participate in a vigorous export trade. He advised crop specialization as the logical strategy for an isolated territory, telling settlers on irrigated land to grow "the things which eastern communities consume, but can never hope to produce, and of which California possesses virtually a monopoly."[100] Smythe recognized the commercial prospects of natural advantages and the power of water to realize them. With the advent of irrigation, California agriculture became intensive. Rather than large farms producing an inexpensive and easily transported commodity, small farms would grow something heavy, valuable, and perishable. Production would no longer expand by breaking fresh ground with old methods; instead, it would increase through reinvestment on existing acres.

By farming according to the natural advantages of climate and soil, California fruit growers hoped to harvest a product of the highest quality for the lowest possible cost. But the balance they sought was a delicate one. Like wheat, any change in circumstance—commercial or natural—might send their new industry into a nose dive. Specialized horticulture required much more capital per acre than wheat did. An orchard represented a tremendous investment to be gambled against the vicissitudes of an untested market and a single perishable crop. High freight rates, saturated demand, insect infestations, and labor shortages might make fruit-growing in California akin to sending wheat from the outermost of Thünen's rings—too expensive to rationally contemplate.

The practice of cultivating a single plant species carried with it a series of consequences that eventually affected almost every aspect of agriculture in California. Over the next fifty years the vulnerability of fruit to natural and humanmade emergencies required a degree of protection that was unlike that which any crop in any region had ever needed. As will become clear in the chapters that follow, natural advantages provided the basis for intensive agriculture, made it a reasonable investment, provided tons and tons of fruit, but said nothing about how to secure the crop against the many threats to its salability. Building California into the greatest orchard in the world meant inventing some institutions and reinvigorating others; it meant bending human beings and nature to meet the expectations of fruit growers and their calculations. It meant a new kind of agriculture, one suited for the world as defined by Oliver Baker and Edwin Nourse.

2

Orchard Capitalists

A landscape of orchards and vineyards took shape in the arid valleys. Though irrigated agriculture fulfilled its promise to create hundreds of densely populated rural townships, smallholdings did not signify that people of modest intentions had replaced the grain kings in the California countryside. Hundreds of square miles of tidy irrigated colonies presented a great contrast to sprawling fields of wheat, but the people who came to take up small tracts differed little from the Chapmans and Friedlanders who preceded them. Horticulture became a business through the efforts of a capitalistic and cosmopolitan group of farmers. The men and women who took up irrigated lands to cultivate trees and vines did not choose to do so because farming was the only life they had ever known, nor because they identified virtue with work close to soil. Fruit growers were more likely to see themselves as business people than as toilers; indeed, the people who settled crop districts from San Jose to San Diego often refused the title "farmer."[1] Instead they referred to themselves as "growers"—orchard capitalists—and they expected more from the fruit business than a bare living. As one promoter put it, they wanted "farming that pays."[2]

The cultivation that the growers favored and the countryside they eventually created reflected this capitalistic impulse. Even with a remarkable reduction in the acreage necessary for a family to make a comfortable living, the self-sufficient farm with its animals, grains, and kitchen garden hardly existed in the irrigated districts.[3] Boosters and land promoters broadcast the astonishing diversity of the state's harvest, including apricots, grapes, lemons, oranges, peaches, pears, plums, and almonds, but cornucopian images obscured the most defining characteristic of California's orchards. Fruit growers often limited their cultivation to a single variety of fruit, a single species of tree or vine, a single product to take to market at the end of a year's investment and labor. As students of land utilization, growers tended to believe that a piece of land had one best use. With the decline of wheat, land monopolists sold some of their acres to another class of rural entrepreneurs to make way for another, more complicated form of specialized farming.

The Irrigated Landscape Takes Shape

From worn-out wheat fields came a new form of real estate. Beginning in the 1870s and increasing in the 1880s, land and water companies built a series of agricultural "colonies" that they sold in ten-, twenty-, or forty-acre tracts. Conjoining land and water required capital. For the few whose land abutted a Sierra river or occupied the moist ground of the Sacramento Delta, water could be had by simple diversion from the main channel or by digging a shallow well.[4] For the many, however, facilities to bring water away from riverbanks proved expensive to build—too expensive, in most cases, for individuals to finance.[5] Individuals rarely possessed the concentrated capital of railroads and banks. Both invested in colony settlement. The Southern Pacific Railroad sold irrigated real estate from portions of its gigantic holdings. The Bank of California held mortgages on vineyard lands in its own colony tract. Financial institutions began to influence economic geography by encouraging one form of cultivation over another. With fruit looking like a sure bet in the 1880s, San Jose bankers loaned up to four hundred dollars an acre on orchards at a time when wheat farmers complained that lenders hesitated to float them fifty dollars an acre—another reason for the decline of wheat. Growers did not settle up the California countryside by themselves. Even before plants took root, fruit needed organization and planning, canals, ditches, and floodgates. The landscape of the industrial countryside was founded by developers.[6]

Developers followed essentially the same process in organizing colonies. First they contracted with an engineer to dig canals and ditches that connected each lot to a major river. The land had to be cleared and leveled. At William Chapman's Central California Colony near Fresno, a corps of engineers and a force of workmen divided each section into thirty-two twenty-acre vineyards, built a system of canals and ditches, and planted grapevines, beginning in 1875 until the winter of 1877, when buyers began to arrive.[7] Other large projects included the Anaheim Colony in Orange County, where Indian and Mexican workers dug 450 miles of subsidiary ditches and 25 miles of feeders for the German immigrants who had purchased the land.[8] The 76 Land and Water Company built a 30-mile-long canal intended to add one hundred thousand acres to the irrigated land of Fresno and Tulare Counties. The entire system, including branch ditches, provided 500 miles of waterways to service two hundred thousand acres. By 1890 thirty-four colonies offered land in Fresno County, with names like the Temperance Colony, the Easterby Rancho, the Washington Colony, and the Bank of California Tract.[9] Small farms in counties across the state provided the basis for a densely settled agricultural society that depended increasingly on the economy of fruit.

The colony became the money-making cousin of the bungalow suburb, and the vineyard and orchard posed themselves as businesslike versions of the gracious gardens that accompanied all respectable California homes. The residents decorated the landscape to reflect financial power and cultured refinement. The Muscatel Estate, located seven miles northwest of Fresno City, featured the intersection of Huntington Avenue and Market Street. Gould, Astor, and Vanderbilt were the other giants of industry whose names distinguished the thoroughfares of the estate's three sections. Fig trees and pampas grass along the roadsides helped to keep the dust down on summer days and gave the residents a shaded place to ride and walk.[10] Simple cottages lined with porches and adorned with ivy, roses, and hedges gave colony homes a feeling of genteel comfort. What the geographer J. Russell Smith would notice about central California in 1925 was also true forty years earlier: "Nowhere in the United States is it more difficult to draw a line between city life, suburban life and country life."[11]

Developers advertised irrigated real estate to people likely to find this blend of rural and suburban living attractive: refugees from urban life and other novices to agriculture. The notion that one could escape from the pressures and dangers of the city without removing oneself from its comforts and rewards became an important attraction of the colonies. Promoters and boosters repeatedly pitched subdivided land to city people. The audience for promotional pamphlets and newspaper advertisements consisted of people ready to take a chance, people who perhaps had failed in other ventures including agriculture, people who could be convinced that fruit growing was easy, profitable, and a fairly certain investment. It did not matter that potential buyers might know nothing about farming at all. Wrote one promoter who knew his audience, "It is to the beginner that I address myself, the man who has engaged in other pursuits; the merchant, the mechanic or the farmer—men whose time has been otherwise employed, and who have had no time to study this business— or, perhaps, the newcomer from the East, who has had no chance to become acquainted with it." [12] For failures in urban life who still managed to maintain an enterprising spirit, colony land provided a dignified escape from public scrutiny.[13] Agents for irrigated land in the Tulare Lake region understood the embarrassment of fallen status, addressing themselves to "business men and society women of San Francisco who are hanging by their eyelids over the ragged edge of business and social anxieties, fretting their souls and gizzards out in expedients to keep up appearances and make both ends meet." [14] A new start with promises of "country ease and independence" and a steady income appealed to insecure urbanites who, though weary, had not lost their material desires.[15]

Growers

Fruit growers mixed back-to-the-land optimism with capitalistic fervor. The new settlers included doctors, engineers, teachers, farmers from the eastern states, and retirees from other occupations, all looking for peace and remuneration. Merchants from eastern cities, immigrants from Europe and Asia, women from Oakland and San Francisco—all came to the irrigated districts with middle-class aspirations for a home and a living. Yet other differences among would-be orchard owners immediately became apparent. People with capital or credit moved more easily into the business than people without. Immigrants from China had few of the opportunities that immigrants from Germany and Italy enjoyed. In an enterprise so costly to establish, investment capital drove a wedge between labor and ownership and introduced class distinctions that characterized the business from the beginning.[16]

The land-hungry people who followed the paper trail to barren lots often had no idea how much money fruit growing required. Since the Revolution, Americans had found land in most places within the reach of the average "actual settler." The land laws too often aided speculators, but tillers found affordable acres in places like Ohio, Illinois, Oregon, Kansas—all along the westward road of Indian dispossession. But in California in the 1890s the cost of starting an irrigated homestead priced out a sizable portion of the population. Small-land farming was expensive to start and sustain. Many would-be fruit growers dreamed of a life in which fig trees and pampas grass adorned shaded walks, but only those with money could afford it.

The first expense was land. During the 1880s colony land (improved and including a water right) suitable for raisins in Fresno and Tulare Counties cost between $40 and $100 per acre.[17] By the early 1890s land values ranged from $50 to $200 per acre, depending on the quality of the soil and the proximity of the lot to transportation and packing.[18] Terms of sale varied. In the 1870s, William Chapman offered twenty-acre lots in the Central California Colony for $1,000 with a $100 down payment and $12.50 a month for six years with no interest. This came to $50 per acre.[19] But the offer of the 76 Land and Water Company was more typical. It included a down payment of 25 percent, with the balance payable over three years at an 8 percent rate of interest.[20] Land at $150 per acre cost $750 up front and then $810 each year for three years. Under these terms, a twenty-acre vineyard summed to $3,180. Highly improved land, which included buildings and bearing vines, sometimes sold as a unit. One mature twenty-acre tract in the environs of Fresno cost $7,000, or $350 per acre.[21]

Land was only the beginning. The vineyardist needed vines (about $400 for twenty acres or about $20 an acre), a house and barn ($1,200), furniture ($500), a well and pump ($100), a wagon and tools ($350), a horse ($200), a cow ($50), trays for drying grapes into raisins (about $174), and packing boxes in which to ship them (about $116). On top of this add the costs of cultivation, including planting the vines, harrowing the furrows, leveling the ground for proper irrigation, pruning until the vines bear, and dusting or spraying to battle insects (about $30 per acre). And as property increased in value with irrigation and improvements, so did local taxes.[22]

The aspiring raisin baron faced a final cost before fruit issued from the earth: supporting farm and family for three years while the vines matured. Vines bear few or no grapes at all the first year and produce at a level below the expenses of cultivation for the next two, depending on local climate.[23] The fruit grower had to "grubstake" by planting temporary crops like vegetables or alfalfa between the rows of young vines as a way to survive the waiting period. As one agricultural economist explained, an investor "should have a considerable sum of money with which to start an irrigated farm, or have the credit with which to meet the production costs . . . which accumulate before he can expect any return."[24] Observers of the raisin business recommended $300 to cover a family's living expenses for a year, in addition to cover crops.[25] All to-gether, assuming that growers used family labor for maintenance and har-vesting in these start-up years, a twenty-acre vineyard might cost $7,500 (or $375 per acre) before the first paying harvest in the third year.[26] Gustav Eisen, whose book about viticulture defined the problems and potential profits of the raisin business, tersely advised new growers to "add liberally to the calculated expenses, if disappointment would be avoided."[27]

Disappointment meant working as seasonal labor on other people's land. Settlers without $300 beyond the down payment could not afford seed for a crop of vegetables. Cows and chickens provided milk and eggs, but the food they furnished or the cash they brought often fell short of the expenses of household and farm. Contemplating future mortgage payments and an unfinished house, struggling growers had two options: labor and speculation. They learned about both from chambers of com-merce and other promoters who strained to put a happy face on the pos-sibility of hardship. "It can readily be seen," went one pamphlet, "that the farmer can cultivate his land during the winter and work out during the summer months at [a] good rate of wages." In a time of labor scar-city and slow real estate sales, developers blurred the line between owner and laborer to attract buyers on the margins of solvency. The pitch con-tinued by revealing the "abundant opportunities for a man's entire fam-ily to be wage earners during the great portion of the summer" and by alluding to the "considerable money" to be earned by following the crops while living in a camp wagon on the side of a road, calling this "an enjoyable outing."[28]

Time spent working for others made it especially difficult to establish an orchard of one's own. A pamphlet of 1887 warned newcomers to tend to their own land. The newcomer could start out by working for neigh-bors now and then, but "a farmer who expects to make a success in farming will find [that a] twenty-acre ranch will take four-fifths of his time."[29] Who would harvest the penniless grower's produce when the entire family went harvesting for someone else? This was not a question that the more established growers in labor-hungry counties cared to an-swer. Poor farmers planted their trees and vines, fell behind in their mort-gage payments, and took to the highways to work alongside Chinese and later Japanese immigrants as migrant labor. Or else they spent decades working for wages close to home.

Charles Carlson struggled in penury and sold his labor in order to plant an orange grove. Born in Sweden, Carlson emigrated first to Illi-nois and then to California, finding his way to the town of Duarte, on the hot uplands between Pasadena and San Bernardino, in 1875. He had designs on ten acres near the city of Orange, fifty miles south, but could not afford to improve them. For three years he worked in Duarte to gather the six hundred dollars that would put him into business for himself. In September of 1878 Carlson visited Orange County for the first time, purchased a pile of lumber, and began to build a house.[30] Without trees in the ground to sustain him, he returned to Duarte. He visited his land once again during the following months to plant and to work for a local grower named L. G. Butler.[31] Carlson lived in Duarte for another two years until March 1880, when he finally moved into his "shanty" and never left. He planted corn for a quick cover crop, but corn could not hold him over.

During the harvest of 1878 Carlson worked for growers named McLoughlin, Lowry (in Duarte), Rodgers, and Tomson. He irrigated, planted potatoes, cultivated orchards, harvested, and drove a wagon.[32] On 24 September 1879 he wrote, "I have worked 5½ days for Mr. Butler this week." The next day he "planted 322 orange trees" on his own land and then irrigated for Butler the day after that in a cycle of work that constantly made him choose between the needs of his own plantation and cash wages. Carlson still worked out in 1882: "Worked for Mr. Stamps[.] $1.50 received from Stamps."[33] Mr. Stamps is mentioned throughout the 1880s as a source of wages. A note at the back of one of Carlson's journals lists Third Street in Los Angeles as an address for Stamps, suggesting that the employer might have been an absentee landowner.[34] Even as late as 1919, though Carlson spent most of his time on his own land and had a family to help him, he still worked out about once a month. Carlson does not fit the image of the grower popularized by boosters and land promoters. His example suggests a class of growers who operated more diversified farms and functioned as a local source of labor for more affluent growers.

There was an easier way to become established. William Smythe, the reformer and irrigationist, wrote in 1905: "As with the miner and the wheat-farmer, so with the fruit-grower the aim was to get rich quickly, and the method speculation."[35] Smythe did not discriminate between growers and speculators—they could be the same person in different circumstances, in different schemes. With minimal capital and fortuitous timing a buyer could purchase irrigated land, improve it over the course of a year while working in some other occupation, and sell it for as much as twice the original price. Thus, although an owner might be financially incapable of improving a piece of property until it paid an annual return, the booming value of irrigated land in the 1880s and early 1890s allowed one to succeed by selling out.[36] It may not be desirable, said one local historian, "that such fertile land be bought for purposes of speculation rather than culture, yet many a dollar has been turned by those who have bought land in the neighborhood described for no other purpose than holding for a rise in price."[37] Two hundred deeds changed hands during the week of 29 April 1887 alone, and the county clerk recorded over $1 million in new property value for the month. The historian called it "speculation running riot."[38]

Here is a dreamlike illustration of the California land bubble from the promotional press. A Mr. Alexander is said to have bought forty acres of unimproved land in the Fresno Colony one winter for $50 an acre, or $2,000. He put in between $500 and $600 in improvements and then sold the land for $3,200 in May to a Mr. Clark. Clark invested an estimated $500 to $1,000 by planting trees and alfalfa on the property then sold it the same September for $5,500. In round numbers, Alexander realized a profit over the cost of land and improvements of $600 in six months (or a 23 percent rate of interest on a total investment of $2,600), and Clark turned $1,300 in five months (a return of 31 percent on $4,200). Speculation suggested a way to build capital without ever attempting to operate an orchard.

The link to speculation is worth considering. Many of those who bought into irrigated lands saw themselves as entrepreneurs first and fundamentally. The notion that intensive horticulture demanded investment far above the cost of land made sense to people used to working with money. A certain portion of the business class in California, the people who cut their teeth on silver and wheat, assumed that they could get rich by rising value alone, without sweaty toil. Gustav Eisen wrote that "many horticulturists expect their fruit to pay whether they care for it intelligently or not."[39] Charles Carlson worked every day for most of his life, but other growers expected not to work at all.

One especially vivid document depicting the life to which many orchardists aspired is *Irrigation at Strawberry Farm* (circa 1888), a painting by Thomas Hill.[40] It is a landscape of the Sacramento Valley. A great Spanish-tiled home presides over a field flooded by two artesian wells. One spills into the furrows to water the delicate crop, while the other serves as a fountain that fills a pond in the shadow of an elegant boathouse. The juxtaposition of two irrigation wells in two very different roles blurs the distinction between business and leisure in the life of the owner who stands between these two landscapes.[41] Wearing a top hat and a long coat, he looks like a cross between an industrial entrepreneur and a southern planter. Under his direction, and bent beneath his gesture, labor five people—including two or three Chinese men—who very clearly belong to the landscape of work and not leisure. The painting emphasizes that the grower is not himself engaged in the cultivation of his field and that the pictured workers are essential. Without them, the owner would either have to harrow and harvest himself or invest his time and money in some other enterprise. Chinese labor made this business suitable to a man in a top hat.

The possibility of picturesque comfort—and not the dreary work of pruning, harvesting, and packing—lured members of the commercial class into orchard ownership. One grower and observer of the industry commented on the laziness of many vineyard owners: "There are, indeed, few raisin vineyards which are properly or even fairly well pruned." The owners, in his estimation, knew little about vines and seemed content to entrust the most important operations to hired hands "without proper supervision." [42] Another critic writing in *Irrigation Age* attacked the slovenly gentleman-grower, saying that he "expects to spend the money while the other fellow does the work." He continued: "Call at his farm and ten to one you find him absent. Either he is on a hunting expedition, or he has gone for the mail. . . . If by chance you find him at home he is either reading the daily papers or smoking a cigar on the piazza. The last thing he ever thinks of is to take off his coat and go to work with his hired men. . . . No wonder he tells you that he can buy butter and vegetables cheaper than he can raise them and that pigs don't pay." [43] Although the author exhorts his readers to tend to their land, it is rare to see a life of hard work advocated in the literature of irrigation or in California's rural press. George Hecke, the first director of the state Department of Agriculture, expressed the work ethic of his neighbors when he said: "No one will dispute that the tendency among men is to accomplish their purpose with as little effort as possible and with the least possible discomfort." [44]

The pace of the fruiting season itself contributed to the notion that specialized horticulture made for an effortless life. Between the spring irrigation and the fall harvest, no major task asked for the grower's attention. The flowers were pollinated, and then came the fruit. So long as the trees had water and sunlight, almost nothing had to be done; and since many growers cultivated only one kind of tree, no other crops created work during the slow times of year. In the fruit grower's ideal this time had no other purpose than pleasure: "During the months of June, July and August, one can go to the sea shore, as a great many do, for there is no care necessary for the vines during that time."[45] Refugees from the scorching Central Valley crowded the beaches at Santa Monica, there to sit on the sand for a weekend or a month while grapes and profits swelled in the midsummer sun. Indeed, this was not like farming as Americans had long known it. The *Overland Monthly* wrote that fruit differed from other forms of agriculture by "an entire absence of the drudgery that we have always heard accompanies a life on the farm; but then, really this is not a farm but a vineyard, and is not to be classed in that same line of effort."[46] The fiction that residents of the irrigated colonies had invented a miraculous tillage that required no labor concealed the drudgery of others from public view.

The existence of laborers to support this life contributed to the reputation of fruit growing as a rich person's business. Few irrigated farmers lived like the well-dressed strawberry magnate depicted in Thomas Hill's painting, but like him they sought to wash their hands of the soil and manage rather then engage in the day-to-day work of an orchard. Promoters drew attention to the laboring population as a way of attracting nonfarmers to colony lands, telling them that owning an orchard did not require the toil of farming.[47] Charles Nordhoff, whose *California for Travelers and Settlers* was one of the most popular books about California to be published in the nineteenth century, told readers that the new settler can "have all his necessary 'improvements' done by contract, even to ploughing his land and putting in the first crop. In this respect labor is admirably organized in California." The Chinese were especially recommended as farm laborers "and are every year more used for this purpose," wrote Nordhoff.[48]

Absentee growers pursued the rewards of the fruit business while shirking off its details. J. W. Pew came to Fresno County in 1883 and started the Mirabelle Vineyard on 160 acres. More impressed with the appreciation of vineyard land than with life in Fresno County, Pew chose to live in San Francisco, where he located his corporate headquarters. Quotidian decisions fell to Mr. Peter Maggette, "foreman in charge," whom Pew called an "accomplished practical vineyardist."[49] William Hall, owner of the 240-acre Montecito Vineyard in the Washington Colony, lived in San Francisco, where he served on the board of police commissioners while his wife looked after the ranch.[50] The Thompson Orchard Company—240 acres of navel oranges and other citrus—had its headquarters at the Mills Building in San Francisco. Harold Burchill superintended the property locally and, quite naturally, had his own 40 acres nearby. Even O. A. Bonnell, the prosperous owner of 13.5 acres, made his home in Los Angeles and paid W. A. Freeman to manage.[51] To generations of Americans agriculture represented the marriage of ownership and work, capital and labor, unified in a single social position. California orchardists attempted to detach one from the other to recast farming as a matter of ownership distinguished from labor.

The *Fresno Expositor* depicted the fruit grower as urbane and progressive and remarkably well rested:

Nothing requires more neatness, more taste, more refinement, or gives more pleasure to the horticulturist than the various labors connected with the vineyard and the packing-house. If we add to this that no other horticultural industry known is equally profitable, and that no other branch of horticulture responds equally to intelligent care and study, it is but natural the refined capitalist, the well-to-do tradesman, the banker, the literary student and the professional teacher, should all look to this industry as one conveniently adapted as a recreation from other work. . . . The raisin vines in Fresno and elsewhere in this State were first planted by these classes of people, and the raisin colony was synonymous with refinement and culture, more so than any other settlement in this State or elsewhere.[52]

In the conception of the booster press, fruit growing was not a sustenance but a venture, not a living but a lifestyle.

Yet, none of these images tell us very much about the kinds of people who became growers, nor do they explain how growers eventually fashioned a powerful industry. We need to know that the indolent grower and the absentee owner existed, but they occupied the margins of the fruit business and were not likely to see their orchards through the price panic or the insect invasion or the labor shortage. The orchardists who organized the fruit trade into California's most important industry before World War II did not spend languid afternoons with cigars on the piazza, nor did they spend years in the employ of others. They came to land ownership with a few dollars in their pockets, joined the chamber of commerce, and used free time in the growing season and after the harvest to look after their greater interests. These growers gathered in statewide conventions, lobbied the legislature, founded cooperative marketing associations, and tested insecticides. They lived in the county, though they may have resided in town, and took an active part in the workings of their property. Perhaps the best way to get a sense of the growers is to look more closely at a few of them.

Minni F. Austin was a schoolteacher who in 1876 poured her savings into the undeveloped Central California Colony. In joint ownership with friends Hatch, Cleveland, and Short, Austin acquired five 20-acre lots and called the property the Hedgerow Vineyard. It is not clear if the other four partners ever moved to Fresno County, but Austin arrived in 1878 and began to grow raisins commercially. In 1885, with her land fully planted and in full bearing, she packed six thousand boxes. In 1886 she packed seventy-five hundred—all from her own vineyard. She was the first grower in the county to own an air-blowing raisin dryer and was among the few large shippers who purchased the raisins of neighboring vineyards to sell under her own label.[53]

A. D. Barling was born in Michigan around 1855, and he later attended Ann Arbor University, graduating with a degree in engineering. Barling moved to Merced County, California, in 1873 to take a position as chief engineer of the Farmers' Canal Company. Eleven years later he traveled to Mexico in the employ of the Mexican Central Railroad. Finally, in 1885, after a term as chief construction engineer for the Southern Pacific's wharf at Alameda, Barling moved to Fresno, where he paid forty-five dollars an acre for unimproved land and opened the El Modelo Vineyard on 280 acres. In 1889, he and his wife purchased a rotary blower to clean and dry their grapes, then built a packinghouse. The Barlings' labels depicted mission bells and rosary beads, images of old California that they hoped would lend romance and stability to the El Modelo product. Demand for their raisins exceeded the size of their crop, and the Barlings began to purchase the raisins of other growers to sell under their own brand. Before long, they employed 450 men, women, and children during packing time.[54]

Ill health brought Erskin Greer across the plains from Illinois to California in 1850. Greer was born in County Tyrone, Ireland, in 1832 and came to Illinois by way of Philadelphia some time after 1836. By the time he turned thirty, Greer had already tried mining at Hangtown, haying near Sacramento, stock raising, and speculating. He purchased his first ranch only to see it destroyed by flood in 1861–62. Greer joined the Sacramento Valley Railroad Company the following year, and while overseeing the road's survey of new townsites began to buy up land for speculation. Almost twenty years later, Greer took the cash he had made from townsites and planted a 100-acre orchard in Sacramento County. Greer served as the superintendent for another orchard and became active in local politics, first as a justice of the peace, then as a member of the board of education and the board of supervisors.[55]

George H. Hecke emigrated from Germany to Woodland, California, in 1891 and immediately purchased 200 acres of orchard land and an equal amount of farmland for dairying. But unlike so many newly arrived landowners, Hecke was no novice to the tree and vine. Born in Hamburg in 1870, he attended the Horticultural College at Geisenheim on Rhine and the Viticultural College at Montpellier, France. After a stint in England with the Royal Botanical Gardens, he made his way to California. At Woodland, Hecke married Elizabeth Welch and went into the business of growing and packing. Hecke quickly turned his attention to building the state's image as a center of fruit production and became an ardent representative of the state's most influential growers. He was appointed to the state Commission of Horticulture in 1916 and later served as the first director of California's Department of Agriculture.[56]

Harriet Williams Russell Strong (1844–1926) grew up near Quincy, California, and married Charles Lyman Strong, a superintendent of the Gould and Curry Mine Company—part of the Comstock Lode region of Nevada. Charles Strong got caught up in the game of chance and invested what must have been a good portion of his capital in a California mine that never hit. Depressed and in debt, he took his own life in February of 1883. While Charles had looked for gold, Harriet spent each year in Oakland and Whittier, south of Los Angeles, where her husband had purchased 322 acres from Pio Pico, the last Mexican governor of California, in 1868. Harriet and her children eventually settled in Whittier, and she went into the orchard business, establishing Ranchito del Fuerte.

Strong immediately faced the problem of having only young trees and what to plant before the first paying yield. "I very soon determined upon walnuts as the chief product, with olives and figs as accessories." As she explained some years later, "[W]hen one has decided that the soil is good for trees, then the problem of planting something to supply interest and tax money while the trees are growing must be solved."[57] Her neighbors planted corn, which yielded fast and adequate returns; but corn also took too much out of the soil, and it shaded young trees and slowed their growth. It was during this time that Strong learned about pampas grass, a bunchgrass native to the plains of Argentina that can grow twelve feet tall. With the critical help of a Chinese hand named Jim, she transplanted a few acres of pampas grass from a neighbor and attempted to sell the plumes for decorative use. She made fifteen hundred dollars that August, then planted 30 acres in 1888, and earned four thousand dollars in 1890—profits that financed the expansion of her orchard.[58] Strong then proceeded to plant walnut trees until she had 150 acres by 1891. The rest of the ranch she put to oranges, figs, and olives.

Strong became active in various aspects of the fruit business in the years that followed. She joined and eventually presided over the Walnut Growers' Association.[59] And though she complained later in life that "a woman needs to have five times as much ability as a man in order to do the same thing," she navigated the politics of the time with remarkable agility.[60] She became active in the Republican Party and managed to convince Mark Hanna, the party's chair, to adopt red, white, and blue pampas plumes as the emblem of the 1896 election. Strong also became the first woman member of the Los Angeles Chamber of Commerce and briefly commanded national attention when she advocated damming the Grand Canyon and selling the electricity to pay for World War I.

But perhaps no grower represented the manner and the ethic of the fruit business as completely as Charles Collins Teague. Born in Caribou, Maine, Teague moved to Kansas with his family in 1881 when he was eight years old. When his father lost the family fortune in a failed property deal, the Teagues left Kansas for Santa Paula, Ventura County, California. Charles Teague went to work for his uncle, Wallace Hardison, and a family friend, Nathan Blanchard, both of whom had been experimenting with lemons. He began his long career in agriculture by pruning windbreaks and picking, but before long Teague took over as labor foreman and joined the management of the company.

The company was the Limoneira Ranch, a lemon venture on 400 acres founded around 1893 by Hardison and Blanchard. By the age of twenty-five Teague had become general manager, and he served as its president from 1917 until well into his seventies.[61] Teague also became the president of two banks in Santa Paula, owned water companies in the area, and eventually directed cultivation on 3,000 acres of walnuts and citrus.[62] Always a player, he helped direct organizations like the California Development Association, the Farmers' Protective League, and the State Horticultural Society. Most significantly, he participated in the founding of the California Fruit Growers' Exchange—one of the largest cooperative marketing associations in the world—eventually serving as its president. Teague stood jaunty and substantial in a photograph taken late in life, with his eyeglasses hanging delicately in his left hand, a watch chain across his chest.[63] Teague represents the apogee of the fruit grower—an investor of capital, an industrialist who knew the land as well as the market, the archetype for a bourgeois planter class that cultivated the land without touching it.

In Teague's conception, agriculture functioned like any other business, with the same decisions and the same risks. This view was not limited to entrepreneurs of his accomplishment. A former dry-goods clerk called farming "a matter of plain business brains, coupled with hard work and the ability to master new problems."[64] The fruit grower without callused hands from years at the plow did not begin at a disadvantage—quite the opposite. His experience in city business freed him from "fossilized traditions."[65] The clerk-turned-orange-grower advised capitalists interested in rural pursuits to "make it a business and run it systematically, as you would any other manufacturing propositions [sic]."[66]

In 1904 the United States Department of Agriculture (USDA) noticed this new farmer in the Far West. The fruit grower cultivated "only one sort or one type of fruit" by employing substantial capital and intensive methods.[67] This "high-class farming" appealed to the "bright and observing" among the population who stood ready to "utilize at all times the results of scientific investigation." The writer identified fruit growers as the leading edge of Country Life ideology—conservationist farmers who carried the urgent message that "year after year the country becomes more thickly settled, land becomes scarcer and more valuable, and intensive methods must gain prominence."[68] To supporters in the USDA, fruit growers used intensive cultivation for rural revitalization and made a big business out of small farms.[69]

Beginning in the 1880s the members of this rural owner class turned their attention to problems of cultivation and transport. Aware that they competed with farmers in other regions in the production of various crops, growers told each other that only "he who can produce cheapest will survive."[70] Cheap production depended, in turn, on the careful manipulation of those natural advantages that everyone acknowledged but that few actually understood. In order for one to produce that delicate "specialty best adapted to his location," he or she needed to know about the ranges and the valleys, about the Pacific High and the rain shadow. The orchardist, in other words, had to know how California functioned as an environment.[71]

Reading the Environment

The landscape that took shape in the irrigated districts reflected the goals of the people who settled there. Nothing about California's natural environment made specialized crops inevitable. Visitors often expressed astonishment at the variety that could be cultivated within the boundaries of a small farm. One orchard, founded in 1853, claimed eighteen thousand trees representing 250 varieties of fruit.[72] On his expedition to California during the Civil War, William Brewer wrote home to New Haven about "the luxuriance of the fruit trees" that appeared like oases in the forlorn landscape.[73] He described a farm in Shasta and one near Mount Hamilton where nature brought a multitude: "This whole valley abounds in the best of fruit: peaches, apples, pears, melons, etc."[74] Yet by 1900, orchardists had begun to carve the state into a series of overlapping districts, where they practiced a degree of specialized farming unheard of in the history of American agriculture outside of the cotton South.

Profit rather than sustenance provided the motive for agricultural specialization. Farmers in California had raised food for cash since they supplied the gold camps, but in 1883 Hugh LaRue, president of the State Agricultural Society, extended principle to common practice when he addressed the members of the annual convention on the flaws of diversified farming. After a survey of the landscape in various districts, LaRue concluded: "I desire to meet the question fairly and candidly, and it is not without due reflection that I say that greatly diversified agriculture is not possible or practicable in our State."[75] LaRue did not mean that no crop but wheat would take root on the flatlands of the San Joaquin, but that no other crop could be grown as *profitably* without irrigation. LaRue supposed for his audience that if an acre planted in wheat yielded ten dollars and if that ten dollars purchased more vegetables than could be grown on the original acre, then the land should be planted in wheat alone.[76] The image of farmers purchasing their vegetables from grocers did not bother LaRue in the least. If the end of agriculture was a money income and not a full pantry, there was no reason to allocate a single square foot to any crop other than the one that paid the best.[77] Irrigation did not change this practice; it simply allowed growers to specialize in other crops.

These other crops proved more fussy than wheat and less likely to pay off in every environment. Wheat kings turned fruit growers learned that the indiscriminate scattering of seed for a fast and dirty dollar did not work with raisins and oranges. Although almost any plant could be grown almost anywhere in California, as William Brewer and countless botanical gardens confirmed, commercial orchardists sought out optimal-yield environments for maximum profit. The economy of fruit impelled this search: the only way for growers to increase profits without cutting expenses was to harvest as much as possible aided only by natural advantages. Yet a landscape that had never known peaches and pears did not announce the ideal location to plant them. Growers selected their plants by reading the published reports of the University of California's Agricultural Experiment Station, by observing the practices of successful growers in their region, and by learning the patterns of a peculiar environment.

Nature, as farmers for generations had experienced it, varied strangely on the Pacific Slope. Tillers from the far side of the Mississippi River found a world without snow or humidity. Leaves still fell off most trees in winter, but orange groves stayed green all year and had their harvest in January and February. The land had almost no humus—the fertile layer of plant and animal debris so prized back East. More bewildering, successive deposits of sand and silt made the subsoils of the Central Valley almost identical in their mineral composition and fertility to the topsoils, even to a depth of six feet.[78] Most important of all, colony settlers had to know exactly what kind of soil greeted the roots they planted. California contained nineteen major soil associations, including more than 150 different local soils and endless variations. It was not unusual to find three distinctive types on one section, especially if the land lay in the floodplain of a Sierra river.[79] "Probably no other part of the continent has so great a variety of soils and products," wrote one visitor, "and both require special modes of treatment."[80]

The maddening diversity of soils reflected similar variations in climate. The most important peculiarity of California's agricultural environment was that it exhibited remarkably different characteristics from place to place, even within a small area. Edward J. Wickson, a professor of horticulture at the University of California and the author of a leading book on fruit growing, cautioned that anyone endeavoring to farm for a living in these arid valleys needed to study local features with a scholar's diligence or risked making costly mistakes. There was, for example, little significance to the terms *northern* and *southern* within the state. Strangers found it difficult to comprehend that "some regions of greatest rainfall have to irrigate most frequently; that some of greatest heat have [the] sharpest valley frosts; that some fruits can be successfully grown through a north and south distance of 500 miles, but can not be successfully carried a few hundred feet of either less or greater elevation." As one reporter put the problem, "The farmer needs to learn his trade over again when he comes to Southern California. . . . 'It is better to come in ignorance,' say many, 'and learn all new than to try to put in practice the principles of agriculture already learned.'"[81] Greenhorns came in ignorance, and they needed help.[82]

Help arrived in 1874 when a German soil scientist named Eugene

Woldemar Hilgard (1833–1916) left Mississippi to accept a position at the University of California. Hilgard believed that a land-grant college owed more to farmers than a four-year education. The university laboratory was their servant and its innovations their property. Farmers who did not demand solutions to the conundrums of cultivation wasted their tax dollars and their energy. The responsibility for finding and disseminating new methods of tillage lay, he asserted, with the state university.[83] Hilgard proposed a vibrant collaboration between the state college and local agricultural societies centered on a series of experimental farms or stations. He recommended that this system, common in Germany, be brought to the United States and that, in like manner, the stations be directed by the agricultural colleges. Mississippi needed more than one or two such farms; in fact it needed "at least half a dozen of them scattered over the State, in the chief agricultural divisions; as many as possible."[84]

Hilgard brought the idea of a proactive agricultural experiment station to the University of California in 1874, founding one of the first in the United States. When, thirteen years later, Congress passed the Hatch Act of 1887 to establish experiment stations nationally, government bureaucrats attempted to influence the direction of Hilgard's station. Washington wanted "pure science" and "new lines" of inquiry at the university—experiment for its own sake. The professor disputed: "I have found in my practical experience that the 'new lines' are most usefully and abundantly suggested by the very work which I defend, viz., encouraging the farmers to submit their practical problems to the Station for solutions."[85] In this way, Hilgard set about to address local concerns with local experiments conducted under the auspices of the University of California.

Hilgard performed the first comprehensive soil survey of California's agricultural regions, relying in almost every instance on farmers who sent him samples for analysis and advice on what to grow. Soil "no. 10," for example, what Hilgard called *sediment soil*, came from the farm of Mr. Daniel Flint on the Sacramento River. He reported of another piece of earth, the gray, gravelly loams found in the San Bernardino Valley: "These lands are said to produce as much as 35 bushels of wheat per acre, and are pre-eminently adapted to fruit growing." Hilgard marked the location of alkali lands, attempted to list the crops known to produce well in each county, and supported the spread of fruit-growing in particular. In one instance he gave his approval to those irrigated twenty-acre lots then just beginning to appear, calling land of that size "the usual homestead units on which industrious families make a fair living."[86]

So much advice so publicly offered had a serious purpose. Long after Hilgard's first study, as competition in the orchard business tightened, the differences between soil groups and slight variations in climate became more critical. C. C. Teague attributed the differing fortunes of lemon growers to "a wide range of soils which affect both the growth and production of the trees and the quality of the fruit."[87] As Teague forewarned his colleagues in 1902: "This lemon business is becoming more and more the survival of the fittest. The poorly located, poorly watered, poorly cared for orchards must inevitably drop out of the running. Southern California is full of orchards which are only monuments . . . [to] the mistakes of those who planted them, [orchards] which never have been and never will be profitable."[88] In 1919 Thomas Forsyth Hunt, dean of the College of Agriculture, advised new settlers that their farmland "is good enough when used for the purpose to which it is adapted, but it is bad when an attempt is made to use it for some other purpose." After a new and more extensive soil survey of the principal agricultural regions conducted between 1900 and 1918, Hunt felt confident enough to inform prospective settlers that land in California could finally be planted with scientific accuracy.[89]

Experienced growers would have accepted Teague's cautions but not Hunt's certainty. Fruit districts emerged as informal subregions—shifting and never well defined—where practice favored the cultivation of one or a few crops as the most prolific products. No one product characterized a crop district, but one may have been most popular. Mistakes did not result in spectacular failure; rather, as Teague suggested, poorly located orchards simply went out of business when their yields no longer supported their costs at prevailing prices. Recall the lessons of the wheat boom: natural advantages are relative, their boundaries and definitions shift with prices and competition and events sometimes having nothing to do with elements as tangible as soil and moisture, but which make the quality and quantity of those elements more vital and thus the location where fruit could be raised at a consistent profit more narrow. Those generalized locations became crop districts.

A rough consensus about crops and conditions emerged along with the spread of irrigated settlement, resulting in a gerrymander of districts all over the state. Pears succeeded in the moist soil along the Sacramento River but could not bear the wet and sticky land along other parts of the delta.[90] Almonds needed deep and well-drained soils like those of northern Contra Costa county, while prunes, plums, and apricots could live with less drainage and thrive in the Santa Clara Valley. Orange trees grew tall and green in the granite soils of San Bernardino but not under the overcast skies of the Ventura coast. Apples did well in high places like the Santa Cruz Mountains because they needed a cold snap once a year.[91] Grapes succeeded in the area around Fresno, halfway between Sacramento and Bakersfield, on the eastern slope of the San Joaquin Valley. Local promoters had reason to boast when they referred to the perfection of Fresno for grape growing. There are only a handful of places on earth where the right climate, water, and soil for successful viticulture exist together, and the southern San Joaquin is one of them.

The raisin district is, first of all, a climate. It only rains ten inches a year in Fresno, with virtually no precipitation at all during the months of June, July, and August.[92] Humidity levels as low as 30 percent compare with those of Death Valley. Temperatures can reach 111°F in the summer, and spring frosts between May and September occur less than one year in ten.[93] Constant heat and no summer rain are preconditions for *Vitis vinifera*—the major grape-producing vine of Europe and Asia. The vine is so sensitive to heat that stages in its seasonal growth are initiated by subtle changes in air temperature. When the average temperature reaches 50°F, shoots (the green yearly vines from which the flowers and the berry sets develop) begin to grow.

This is the time for irrigation. According to Frank Adams of the University of California, deciduous fruits like grapevines need between one and three six-inch to nine-inch irrigations a season to ensure adequate wood and bud growth and to develop the fruit.[94] Adams recommended 1.5 acre-feet of water per acre per year for best results, in addition to twelve to eighteen inches of rain.[95] To illustrate this quantity, if all the water needed to cultivate a twenty-acre vineyard for one year could be poured onto a single isolated acre it would flood the land fifty feet deep. Properly irrigated soil and the onset of spring cause grape shoots to elongate as much as an inch a day. At 68°F, usually recorded during the first week of May in Fresno, the first flowers open, signaling the beginning of the fruiting season.

Water and soil work in conjunction to sustain the plant. Irrigation not only makes possible the transportation of water, but its measured application. Too much moisture can be as damaging as too little in the reproductive life of a grapevine. During the fruiting season, extra water in the form of unwanted irrigation or rain will damage the developing grapes. Vines use water for vegetative growth any time of year, and vegetative growth in the last month before the harvest competes with the developing fruit for the plant's store of carbohydrates. Grapes with a low sugar content result.[96] How well vines absorb water has to do with the quality of the soil. Prime vineyard soil holds water long enough for the roots to take it in, yet is porous and deep enough to trap air, drain well, and allow extensive root development. The stubby bunchgrass and tall tules that covered the valley in advance of wheat fields grew in this deep, silty earth, the result of centuries of alluvial deposits from Sierra rivers. Reddish loam, white ash, and "sand hill" are a few of the more celebrated loams of the San Joaquin Valley.[97] All together, intense sunlight, regulated water, and porous soils promote fast, vigorous growth and exceptionally sweet grapes.[98] Few crop plants thrive in such conditions, but eight-year-old muscat vines can produce ten to thirteen tons of grapes per acre in a season.[99]

The fortunate combination of plants with soil and water produced great yields in other districts and began to define a regional agriculture that set California apart. Apple trees produced up to eight tons per acre.[100] Pear trees brought their owners between fifteen and eighteen tons per acre. Peaches and apricots yielded five to nine tons per acre. The intense heat of the growing season matured the fruit so quickly that the first harvest could be called in late June and early July for many deciduous fruits.[101] Gustav Eisen reminded vineyardists of the economic advantages that accrued from natural advantages. He said that if raisin grapes could be grown and cured anywhere at all they would be worth no more than vegetables: "The safeguard of the raisin industry is therefore the limit to the conditions under which the grapes can be grown and under which the raisins may be cured."[102] Even with fruits like peaches and pears, over which California had no monopoly, the product ripened so early and with such an excellent flavor that they too became distinctive to California.[103]

Relative Advantages: New York's Frost

The test of California's natural advantages was how they compared to those of other regions, some of which had supplied American cities with fresh fruit for more than a century. Michigan and Wisconsin had their lake districts where the shore influenced a climate temperate enough for peaches. But no state compared to New York in its importance to American horticulture. New York, New Jersey, and Delaware composed the fruit and vegetable capital of the continent in 1900. New

York's advantage came from its proximity to Manhattan, Long Island, and southern Connecticut—a concentrated urban population that in 1900 included 10,060,983 people or 13 percent of the United States.[104]

Long before fruit began to arrive from the Far West, New Yorkers grew it locally.[105] It became something of a craze for middle-class people to cultivate small fruits and vegetables in lean-to greenhouses. Hothouse farmers supplied their plants with extra heat in winter by running the smoke from the house through pipes along the inside of the glass. Apples, pears, peaches, apricots, cherries, nectarines, strawberries, and pineapples could be raised through a practice called "forcing," a term that suggests the degree of human intervention necessary to coax plants in a dank and clammy room into reproducing. Until the 1880s profits justified the considerable expense of this cultivation. Ulysis P. Hedrick, historian of agriculture in New York, reported returns from hothouse grapes of as much as six hundred to a thousand dollars a year.[106]

The fruit industry in New York, however, did not arise from modest hothouses but from orchards. In 1915, Hedrick described the environmental basis for commercial horticulture in terms that any contemporary Californian would have understood. He called New York's climate and soil "so diverse and so favorable as to make possible highly specialized pomological areas" and suggested that farmers learn the boundaries of soil regions so that they may "specialize more closely, for each fruit and each variety of fruit has a set of conditions best suited to it."[107] Famous crop districts included the Hudson River Valley, Long Island, and the St. Lawrence and Champlain Valleys.

Fruit grew to great quality in New York, even with some serious climatic limitations. Regarding the St. Lawrence Valley in particular, Hedrick wrote: "It is hardly necessary to say that only the hardiest varieties of the hardy fruits thrive in the uplands, and that only in favored locations near the water can any of the tender sorts be grown." Farmers in this location lived one cold wind away from ruin. As for fruits planted in the upper Mohawk Valley, "hardiness is a prime requisite" here too. The grape region to compare with Fresno, the Chautauqua grape belt, extended along the Erie Shore from the Niagara River to the northwestern corner of Pennsylvania and contained thirty-five thousand acres of native Concord vines.[108] When Hedrick described this region as "temperamental" he probably referred to the disastrous frost that seized Chautauqua in May of 1895. The freeze hit at the peak of the fruiting season and sent scientists from Cornell University hurrying to determine which areas had survived.[109]

By the first decade of the twentieth century, general agriculture in

New York had suffered a serious decline. What Liberty Hyde Bailey observed on his tours of the countryside was substantiated by other studies. Ralph Tarr, also of Cornell University, stated that "the most basal cause for the decline is the competition of products from other agricultural regions." He had in mind grain from the Mississippi Valley, not pears from Sacramento, as the reason for abandoned farms and vacant schools, but his conclusions are telling. Tarr said of the farmers in his state that they had failed to adapt themselves to changing market conditions "and to modify the farm products accordingly." They labored with soil called "thin, stony, and naturally infertile" that no longer provided yields sufficient to justify the effort of tilling it. "Extensive farming is the rule," wrote Tarr, "and of intensive farming there is rarely any evidence." [110] New York fruit growers may have enjoyed the advantage of good location, but their natural disadvantages and antiquated methods provided an opportunity that Californians exploited.

Orchardists in California heard about frost in May and other afflictions known to the East and trumpeted their own natural advantages as valuable "beyond computation." [111] Without knowing very much about other fruit-producing regions, they disdained any attempt to grow fruit east of the Rockies. One gloating member of the fruit convention remarked that trees in many eastern states took eight to ten years to come into bearing and that their owners lived in constant fear of frost. After telling about a man from Ohio whose peach orchard perished in a snowdrift, the grower crowed: "Can peaches, pears, plums, prunes, and apricots be produced in paying quantities there? No; not at all." [112] N. P. Chipman quoted selectively from the proceedings of the American Horticultural Society when he described the "discouragement and disaster" of fruit growing in the Middle West, taking special note of the wilting droughts, violent winds, and brutal cold. To resist the devastation of frost in late spring, one pitiable Kentucky farmer proposed to place coal-burning stoves between every four trees and to light them at first news of a threatening wind. Chipman's rejoinder emphasized the high price of bad weather: "Imagine Mr. A. T. Hatch . . . with his orchard of six hundred acres, putting stoves in each square of four trees—twenty-five stoves to the acre—fifteen thousand stoves to fire up and keep going!" California growers took such examples as evidence of the moribund state of commercial horticulture in the other sections and believed that the great eastern cities offered them an untapped market. [113]

The state Board of Trade concluded from this and other evidence that California was in the process of becoming the nation's fruit basket. Year to year, the growth of the business astonished observers and attracted attention to the Far West. In 1882 Fresno County shipped 80,000 pounds of raisins; in 1891 it shipped 44,954,850 pounds. By 1898 the state produced and shipped 80,000,000 pounds.[114] Fruit exported from the port of San Francisco arrived in Australia, Amsterdam, Hawaii, Japan, Bombay, and Russia.[115] "Fresno raisins," one promoter declared, "are already famous in Europe."[116] By 1924–25, the district supplied 92.4 percent of the grape tonnage shipped by rail in the United States and produced 90 percent of the raisins Americans consumed—all grown in an area about the size of the state of Delaware.[117] California shipped 302,587,970 pounds of fresh, dried, and citrus fruit by rail in 1891, an amount equivalent to 15,129 carloads. One year later, the state sent an additional 60,368,965 pounds, representing a 20 percent increase. Shipments of green deciduous fruit alone increased almost 50 percent in one year.[118] Yet numbers like these conceal a volatile and unpredictable commerce in which orchards often lost money. Realizing the benefits of ideal environment meant attuning the land to the market, a process that began when growers adapted their businesses to the sale as well as to the production of fruit.

Economies of the Packinghouse

To make the orchard business pay, growers had to master the distant market as well as the local environment. Throughout the nineteenth century American farmers sold an increasing portion of their surplus for cash in cities, but few of them limited their tillage to a single money-making crop, producing little or nothing for home consumption.[119] Farmers who rested their fortunes on one crop relied on propitious returns for the very bread upon their tables. They flaunted their good fortune before the old vagaries that plague the farmer: bad weather, bad prices, and bad luck. Though the land produced an almost constant harvest, without an unobstructed path to consumers all that plenty might go to seed.

Intensive farming meant spending more money on limited land to promote significantly higher yields, but it said nothing about how to *dispose* of those yields. During the days of the boom, when fruit in almost any condition sold high, many orchardists failed to recognize how much they had in common with other farmers. Like other farmers, they depended on city people to purchase their commodities. Land promoters wrote at length about the large and lovely fruit that irrigation made, but they left settlers to figure out how to transport the product and arrange for sale. Before the many members of the industry federated to establish cooperatives and a coordinated system of commerce, individual growers attempted make market connections for themselves. Orchard owners hit upon a key insight into the nature of the fruit trade that would influence the entire relationship between rural California and the people of North America: there was more money to be made in the distribution of fruit than in its production.

Since the Gold Rush, farmers in California supplied urban populations. Chinese vegetable gardeners in Stockton made more money selling onions and cabbage to malnourished miners in Placerville than by mining.[120] Farms in the Sierra foothills, on the Sacramento River, and in the Sonoma and Napa Valleys—some as large as six hundred acres—sent grains, livestock, fruits, and vegetables to San Francisco and Sacramento, vending directly to grocers or jobbers.[121] It did not take long for increasing production to satiate local demand, and soon after the Southern Pacific and Union Pacific railroads connected the coasts California agriculture came to depend on more distant cities. Meeting in 1884 at one of the first annual statewide horticultural conventions, growers agreed that making a good living meant selling beyond the state's borders: "We must have an outlet for our fruit. In ten years from now this State will raise fifty times more fruit than is needed for home consumption and the Eastern market must be opened for us."[122] They regarded such a trade as the very salvation of their business in bad times and the basis of its future stability.

Far-flung commerce carried enormous risks. Growers bet windfall against ruin by sending uninsured crops representing a year's income in freight trains and ship bottoms to buyers thousands of miles away.[123] In 1893 one syndicate sent a box car of oranges to New York, packed it into a steamer, and dispatched the load to London to prove that they could open Europe to California crops. The growers furnished the fruit and the shippers absorbed the transportation costs. "There seems to be no possibility of loss, and every probability of gain," said one of the partners[;] "it is simply a question of amount of profit."[124] The consignment took about three weeks to reach London (fifteen days to cross the Atlantic), a long time for oranges to sit unsold. But the seas and the fates brought $3.50 per box, money split between the English consignors, the shippers, and the growers, who made between $1.00 and $1.50 per box.[125] Five years later, a San Francisco firm sold $400,000 worth of dried fruit to merchants in Hamburg, Germany. It was the largest contract of its kind at that time.[126] Such ventures appealed to those who wanted to make a quick fortune on a onetime deal.

Most orchardists tried to open up regular channels of commerce, only to find that dealing with wholesalers in American cities could be as difficult as selling to the ends of the earth. George Kellogg's attempt to sell from his orchard in Placer County to the merchant houses in Chicago and New York illustrates some of these difficulties. Kellogg published a book of private telegraph codes in 1893 to facilitate wire communications with his distributors. Any merchant or broker who wanted to do business with Kellogg would have kept this book by the office telegraph and used it to send orders and requests to California. The cipher lists thousands of words, each corresponding to a sentence or phrase. For example, next to "Acute" we read "Have received remittance," "Bed" meant "Ship immediately," and "Baptism" over the telegraph informed the merchant on the other end that Kellogg "Will do our best to ship today (not a promise)."[127] But other of Kellogg's code words suggest that transcontinental trade sometimes went wrong: "We fear telegrams unsafe except in cipher"; "We think something is wrong among RR agents"; "Will use best judgment in selecting, packing and shipping goods, after which it is understood you assume all risk."[128] It would be years before regular channels functioned.

Nonetheless, when California fruit arrived in Chicago and New York in salable condition it made its owners extraordinary profits. The difficulty of finding buyers and sending fruit actually made it more valuable. The *New York Sun* reported that "the prices which are paid for fruit thus sold would make a domestic producer's head swim." Hatch-brand Bartlett pears, "the best sent from California," sold in boxes equal to about one-third of a barrel—the basic unit of the eastern trade. The pears were so large that no more than fifty of them fit into a standard box. One week in 1889 the trade sold 2,800 boxes of California Bartletts for an average price of $3.75 per box, while New York–grown pears failed to find buyers at any price over $3.50 per *barrel*.[129] At the price paid for a box of California Bartletts a New York barrel would have sold for $11.25. Grapes told a similar story, with western shipments finding as much as $7.50 for a forty-pound case (about $.19 per pound). Grocers considered them a delicacy that only sold, in the words of one merchant, to "high-class" buyers and "first-class places."[130] California grapes retailed like imported caviar while the local Concord grapes sold for less than $.04 a pound or from pushcart vendors for $.03. In the three years between 1886 and 1889, the trade in California fruits increased ten times, prompting a New York reporter to remark that fruit from the Pacific Slope was "competing with the domestic fruit product, and beating it out of its boots." [131]

Raisins exemplified the copious yields and high profits that colored the early years of the business. A typical twenty-acre vineyard with five-year-old vines yielded 5 to 6 tons of grapes per acre per year.[132] Grapes have three times the water and thus three times the weight of raisins, so after two weeks in the sun the grower collected about 1.8 tons per acre to send to the packinghouse. The basic unit of the raisin trade was the twenty-pound box. One and eight-tenths tons per acre equals about 180 boxes per acre, or a total of 3,600 twenty-pound boxes from twenty acres. Good prices for a box ranged from $3.00 in 1873 to $2.00 before the depression of 1893.[133] With the somewhat lower price of $1.75, the owner of this typical vineyard would gross $6,300 (or $315 per acre).[134] Subtracting the costs of cultivation, labor, taxes, insecticide, and marketing (about $1,800 total), the owner could expect a net profit of $4,500 for a season (or $225 per acre, $125 per ton).[135] For orchardists with a total invested capital per acre of $400, $225 represented a return of 56 percent, and the vines yielded an extra ton per acre the next year and still more the year after that.[136]

Proponents of intensive fruit growing loved to list the enticing returns of an acre "under the ditch." One author claimed without exaggeration that irrigated fruit returned between $76 and $232 per acre.[137] Gustav Eisen called profits of $200 to $250 per acre common on the smaller vineyards. The average returns on larger vineyards reached $125 to $150 per acre. As much as $2,900 was recorded as the gross return from twenty acres of *three-year-old* vines. "The entire country is growing into one solid mass of settlements, of five, ten, and twenty acre lots," reported one chronicler of Fresno County, "on which the owners are not only making a comfortable living, but many of them are laying up what an Eastern farmer would consider quite a fortune, many netting from $4,000 to $7,000 annually." [138] Fabled stories of vineyards and orchards that brought their owners $480 *per acre* for raisins, $800 for currants, $1,500 for Bartlett pears, and $2,000 for oranges became common stock in the trade press.[139]

Possibilities like these led to new planting. The prospect of riches from prunes and pears sustained a migration of investors into the irrigated districts. Nurseries in Southern California reported record sales in January of 1893. Thousands of acres of wheat and barley turned into olives and lemons within weeks. A visitor from Boston reported to the *Pacific Rural Press* that eastern cities remained undersold and understocked, that no place in the country had trees of the same quality as those in California, that overproduction was "an absurdity." [140] In about ten years, from 1883 to 1893, the state added three hundred thousand acres of fruit trees, and planting "on an enormous scale" was said to be in progress. [141]

The drive to win money in the market caused another kind of expansion, one very important to the future organization of the industry. Rather than buy more land, some growers acquired packing sheds and lithographed labels and assembled many small crops into large shipments. Their motivation had to do with attaining economies of scale, which result when an increase in the size of a manufacturing enterprise decreases the cost to produce and distribute each unit. Nothing is more important to the creation of economies of scale in production than labor-saving machinery—an element notably absent in the fruit business. [142] Without machines to replace hand labor in the harvesting and pruning of orchards, the owners of two hundred acres enjoyed almost no cost advantage over their twenty-acre neighbors. [143] As for irrigation, when gravity and not electric pumps fed water to the land, economies of scale did not exist here either. Water cost the same per unit no matter how much of it one purchased. And intensive methods tended to work better on smaller orchards than on larger ones. [144] Groves the size of sections certainly existed, but they depended on expensive management to keep them profitable.

Marketing offered economies that production did not. The packing-house suggested a way to reduce the cost of production by selling as many boxes as possible, including the crops of neighboring growers. One observer of the raisin industry described the process by which a grower with a packinghouse became a shipper: "Raisin making is so profitable that many of the owners of vineyards have orders for their brands in excess of what they produce; and to meet this demand they purchase the grapes from neighboring vineyards at $20 per ton on the vines, the purchaser being at all expense of picking." [145] Shippers attained a measure of flexibility in supply without assuming the cost of improving and operating additional land. In time, packing companies produced only a fraction of what they sold, choosing instead to serve as buyers and commission houses for the hundreds of small growers who found it nearly impossible to make their own arrangements with wholesalers. The packing shed stood in the borderland between production and marketing, and growers who could find their way on both sides of the line became denizens of two worlds, one composed of cultivators and the other of the people who ushered crops to consumers.

The company of R. B. Blowers and Sons illustrates this trend. The family-owned company raised grapes for fresh sale and raisins on eighty acres near the town of Woodland in Yolo County.[146] Early in the 1880s R. B. Blowers began to send their grapes to New York in refrigerated and ventilated railroad cars to the merchant house of Sgobel and Day. The merchants found abundant buyers, and demand for grapes from the Blowerses' vineyard soon surpassed the supply. Rather than report the end of their stock and close for the season, the vineyard company began to consign the grapes of neighboring vineyards under the same label.[147] This practice expanded the volume of their sales and helped them to pay for the cost of labor in their vineyard and packinghouse. R. B. Blowers soon attracted solicitations from brokers who wanted a piece of their business. One salesman called the company "one of the principal raisin makers," and asked to represent them in "the South." The South, in this case, was Belize and Honduras.[148]

Another firm, the Earl Fruit Company, began as a subsidiary of the Armour Packing Company of Chicago. The only companies that owned refrigerated railroad cars in the 1880s were meat packers like Armour and Swift, and when they sold beef to California they needed some commodity to fill their cars on the return trip. In order to assemble a large variety and quantity of fruit for this purpose, Armour established the Earl Fruit Company, managed by Edwin T. Earl. Like Blowers, Armour built a packinghouse (eventually many of them throughout the Sacramento Valley) and charged local growers for railroad freight and refrigeration to Chicago. According to Joseph DiGiorgio, whose family purchased the Earl Fruit Company from Armour in 1910, "the Earl Fruit Company became the dominant commercial packing company in California and really controlled a great part of the business."[149] DiGiorgio came to Baltimore from Sicily in 1889 and took a job in a wholesale fruit warehouse. By the age of thirty-six, he had established the Baltimore and Atlantic Fruit Company, a corporation valued at $12 million in 1911— the year he moved to California. The family entered the fruit business through distribution, not cultivation, but the DiGiorgio Fruit Corporation eventually controlled over twenty thousand acres in California and Florida.[150]

The purpose of these examples is to demonstrate that one way small scale became large scale was through distribution. By the 1890s the fruit districts included orchard and vineyard companies considerably larger than twenty acres. The Butler Vineyard in Fresno County included fourteen hundred acres with four hundred acres in raisin grapes. Butler shipped six hundred tons representing 40,240 boxes in 1886, all under his own brand.[151] A. D. Barling shipped enough raisins in a single year to fill eighty-three railroad cars.[152] G. W. Meade and Company sold 75,000 boxes of raisins and the Curtis Fruit and Packing Company sold 22,432.[153]

Yet the affluent society of the colonies was more precarious than it appeared. By the middle of the 1880s, growers in specialized districts had discovered that raising a single crop carried all kinds of unforeseen hazards. They found that California's six or ten important fruits could be imperiled by any number of emergencies because they shared the same vulnerabilities. "With this intensive farming," opined a speaker before the Agricultural Society, "diversity of product must go hand in hand, for it is a dangerous condition to have but one product on a small farm. In good and in average seasons the crop will pay wonderfully well, but then there comes a bad season, a poor market, *or any of the many conditions that are adverse*, the small farmer finds himself without means of sustenance. . . . "[154] After the first flush years, after the land bubble burst with the depression of 1893, the productive power of the land ran up against a market incapable of absorbing its bounty. This time of "readjustment to new market conditions," wrote one irrigation expert, was when the "real and discouraging hardships of the new settler are felt, the time when water payments lapse and the less persevering give up the struggle."[155] As long as the breathless balance between the fecundity of the soil and the viscosity of demand maintained, the crop supported the people who grew it, but as we shall see, prosperity in the fruit business was by no means as predictable as the rush of cold water from an open floodgate.

Diversification as a way to spread risk and resources over a number of crops carried little force among growers. It did not appeal to people who saw themselves as manufacturers and whose land, irrigation systems, tools, plants, and knowledge had all been acquired with one crop in mind. As vulnerable as it made them, growers considered specialization their only device to influence the market. George Roeding, a grower who popularized new varieties and promoted commercial horticulture, explained the quandary of the specialized orchardist. Speaking before the State Agricultural Society, Roeding argued that diversification represented one kind of financial safety. Farms in the Middle West were large and diversified he said, "so that in the event [that] the price on one [crop] is low, due to over production or some other cause, the others will counter balance any loss." [156]

But the producer who diversified might be rendered powerless in a market that required one to command as great a portion of it as possible. Roeding reasoned that "the only successful method of growing fruit in California is to have enough of any one product to make the seller a factor in the market." "In other words," he continued, "a man who devotes, say[,] twenty acres to twenty different varieties of fruits, will never be in a position to secure as much for his product as the grower who devotes his energies to a few products." Possessed of this logic, the owners of the California countryside pursued market power at the expense of social and environmental stability.

With all its risks and costly tendencies, specialization went unchallenged and virtually unquestioned, even as rural California suffered economic depression. Instead, orchard owners and their advocates demanded protection for the practice of monoculture. Rather than diversify to earn for themselves a small degree of self-sufficiency in chaotic times, orchardists chose, instead, to fortify specialization against such formidable disadvantages as long-distance trade, perishable products, the ravages of insects, and an insufficient labor supply. Indeed, there is no adequate environmental explanation for the emergence of industrial horticulture in California. Natural advantages assured orchard owners of only one thing: that they would harvest great quantities of fruit for their capital and labor. Bringing fruit to the continent required much more than land and good climate.

3

Organize and Advertise

The land was advantageous, but its location was not. Though blessed with the best soil and climate for their particular crops, fruit growers had to contend with one overwhelming and unavoidable fact: their geographical isolation from the largest cities in North America. The Sierra Nevada, the Great Basin, the Rockies, the Great Plains, the Mississippi and Ohio River Valleys, and the Appalachian Mountains stood between California and the great centers of consumption. Fruit growers remained stubborn adherents of David Ricardo in a world haunted by Johann Von Thünen. No other region could produce raisins and oranges as fine as those grown in California, but commerce depended on a marketplace that did not cost more to reach than the value of the products sent.

Turgid speeches about the "Garden of the World" and its "physical contiguity" to Chicago and New York did not alter geographical reality. Although the world had indeed become a smaller place, railroads and telegraphs only made the idea of sending perishable fruits over long distances more plausible than it had seemed before 1869. Though boosters wished otherwise, Chicago had not moved from its position at the southern tip of Lake Michigan and New York City remained an Atlantic port. Distance had hardly ceased to be a factor, and communication with merchants in the two cities still caused confusion on both ends of the line. When fruit began to pile up in orchards, rot in transit, and sit unsold on the docks at Manhattan, "natural advantages" seemed irrelevant.

Growers saw more peaches and plums hanging heavy on the bough than they knew could be handled by existing methods of distribution. They further understood that in order for commercial horticulture to survive its own expansion, its producers needed to organize and reach out to consumers in a manner similar to manufacturing corporations. Orchardists had anticipated the end of their isolation ever since the Central Pacific Railroad linked Sacramento to Omaha. Though a necessary condition for California's expansion, the railroad was not sufficient. Growers could send their fruit anywhere, but they often struggled to sell it because they depended on marketing agents and institutions that repeatedly failed them. As passive producers scattered over a remote territory, they could neither regulate supply nor influence demand. Orchardists needed an organizational form that expressed their individual interests and embodied their common needs, and the agricultural cooperative marketing association provided that form.

Selling fruit as a consumer product forced a change, not only in the structure of the trade, but in the product itself. A soft ripe pear fulfilled the tree's purpose by nurturing a seed for germination, but it did not fulfill the purpose of a mass-production industry. That pear could never be allowed to express its inherent purpose. It had to be picked, cleaned, sorted, packaged, examined, shipped, unloaded, stored, and boosted to the public as a new necessity of urban existence. It had to withstand passage through many hands, sit in a boxcar for over a week, and still look fresh in the grocery store window. In time, the landscape of rural California came to reflect the exacting standards that cooperatives set for fruit and the mercurial preferences of shoppers. As it opened new pathways of consumption the cooperative also interjected the California countryside into the churning center of the American economy.

When Merchants Ruled the Trade

The orchard business withered in its isolation from the wider world. In 1909 fruit generated more money than any other form of agriculture in California, but it had already hit a wall composed of ill-suited methods of transportation, cunning distributors, and an uninformed public.[1] Speakers at farmers' conventions and county fairs described waste and disorder in the trade and called for reform. One of those who came forward with a series of ideas for how to bring fruit from the Far West to the urban millions was a Jewish progressive named Harris Weinstock.

Weinstock spent forty years of his life navigating and formalizing the relationship between California fruit growers and the markets for their produce. His movements and projects provide a device, a motif, an entrance into the tangled subject of markets. Weinstock is certainly not the whole story, but his early career spans from the 1880s to the second decade of the twentieth century, from the founding of the California Fruit Union in 1885 to his term as the first state market director beginning in 1916—the critical period of advocacy and formation during which California agriculture entered the national economy.

Like others who sought their fortunes in the fruit business, Weinstock owed nothing to a rural childhood. He was born in London to Solomon Weinstock and Rachel Lubin in 1854 and moved with his family to New York a year later. In 1869 Weinstock joined his half-brother David Lubin in Sacramento, where the two entered business together as general merchants, selling tools and equipment to farmers. The venture led them to purchase land of their own, and in 1884 the brothers acquired twelve hundred acres, including a vineyard, in Colusa County. Now a fruit grower, Weinstock learned the difference between selling a sack of flour from a storefront in Sacramento and a box of grapes through a commission merchant in Chicago. Selling anything through unknown receivers made for risk, but when the commodity in question had a tendency to decompose between point of production and point of sale, it made for ruin. Turmoil in the fruit trade offered Weinstock his very own progressive mission, a chance to establish process and banish corruption. Not long after acquiring his vineyard, Weinstock took up the task of making the world safe for entrepreneurs like himself who depended on a single money-making crop.

Specifically, Weinstock set out to improve the poor state of marketing. In his first effort he helped to found the California Fruit Union to dispatch the harvest of many growers through one commercial channel in an attempt to shift the burden of competition from producers to buyers. In this way, stated Weinstock, "the profits, which formerly had gone to shippers and middlemen, would be made by the grower himself."[2] The Fruit Union would fight for lower freight rates and prompt returns. The organization also promised its members information about supply and demand in destination markets, a hard commodity to come by on the far end of the tracks.[3] The fact was that the members of the Fruit Union knew little or nothing about what actually happened to their shipments once boxcars pulled away from Sacramento and even less about what befell them in Chicago and beyond. So, as an attempt to gather intelligence about subjects that had remained mysterious for too long, Weinstock boarded a train for Chicago and points east in 1886. What he found while on reconnaissance in wholesale markets and avenue groceries convinced him that only an organized effort on the part of growers could change a system that so obviously worked to their disadvantage.

The first problems that Weinstock found had to do with the physic of fruit, its form and tendencies. Fruit perished a short time after harvest. It ripened, softened, and began to decompose, using its sugars to nourish the seed. No other problem so bewildered and frustrated those who endeavored to sell fruit as the fact that it could not be stored and transported with any certainty. Like most commodities in the nineteenth century, fruit went to market on the railroads. But consider the differences between fruit and one of the most common agricultural products at the time. Grain could be stored for long periods in elevators and boxcars with no degradation. It simply required a cool dry place. The hardiness of wheat and corn influenced the way people bought and sold them, making possible a new kind of market in which elevator receipts representing stored grain traded on the floor of the Chicago Board of Trade.[4]

Fruit could not be sold with the same confidence. When properly cooled, a bunch of grapes lasted about ten days between vine and market basket before it molded. With the refrigeration technology of a century ago, an unripened pear had as long as three weeks if kept cold. Peaches and plums had less time to find a buyer. Oranges, with a higher acid content, had more. But no matter the variety, if it was not moved from west to east within days, a piece of fruit lost all of its value. Any bruises sustained during packing hastened decomposition; any insects affecting a single box might spread to an entire carload during transportation—even in a refrigerated railroad car.[5] Once out of cold storage, fruit ripened in a matter of hours. All through the 1880s and 1890s growers and shippers observed that while their volumes increased, their profits diminished. They identified many problems with the business, but perishability, transportation, and the conditions of sale landed at the top of the list.

Fruit required speed and good storage on its way to market, but the railroads provided neither. For one thing, railroads did not modify their schedules to suit the needs of this delicate cargo. A shipment might arrive at the point of sale in six days or in sixteen depending on how the railroads directed their cars.[6] The long haul had more to do with railroad economics than it did with the distance between major points. Boxcars heading west out of Chicago often traveled empty because the factory-made goods they carried took up less space than the grain, livestock, and produce they had been sent to collect.[7] To recover the cost of sending a train all the way to California, eastbound railroads made many stops through the deserts, mountains, and plains to pick up as much freight as possible on their way back to Chicago.[8] This made estimating the time required for a carload of fruit to reach its destination like throwing darts. One economist recommended that "it is now necessary to discount the scheduled time of arrival, and to base diversions largely upon guesswork."[9]

Refrigerated railroad cars did not solve the scheduling problem. Although ice made it possible to transport perishable commodities two thousand miles, it could not compensate for a trip as long as thirteen days to Chicago and nineteen days to New York.[10] If the destination city happened to be glutted with too many peaches and pears on the day of arrival, keeping a carload in cold storage for five more days until prices improved meant risking that the fruit would fall apart upon first inspection. Still, refrigeration offered a better risk than the alternatives. In the sorry event that companies like the Fruit Growers Express had no cars available for a shipment ready to move, the crates traveled in ventilated freight cars or piled next to luggage, with predictable results. By the time they reached New York's Terminal Market, where the majority of California produce landed, grapes and pears had to be rushed to the auction room and sold before buyers could discern any mold or browning.[11] Harris Weinstock spoke to one New York agent who bluntly told him, "You must remember that thus far we have not been able to bring carloads of California fruit to New York in good condition. There has always been a large percentage of loss by decay while in transit."[12] The stuff did not sell well because by the time it arrived no one wanted it.

California growers wanted six days to Chicago and eight days to New York, a schedule intended to bring fruit to new populations. Representatives of the industry argued that if fruit arrived at major points in good condition it could be "forwarded to other locations from two hundred to five hundred miles distant."[13] This plan would feed millions of new fruit consumers, people "who before saw but little, if any, of it."[14] In other words, fruit made the best money when it could be rehandled for dispersal to smaller cities—such as Madison from Chicago and New Haven from New York—where it fetched higher prices.[15] Weinstock discovered that grocers in Washington, D.C., had seen scarcely fifty crates of grapes from California; in Baltimore they had received only seventy.[16] When retailers in Pittsburgh ordered California grapes from wholesalers in New York, they paid high prices "and received stuff that proved almost worthless."[17] These cities did not lack California fruit because buyers did not know about it. Poor transportation made them hesitant to order it.

If they survived the passage over mountains and plains, apples and oranges from the Pacific began another, more complicated journey through the offices of the assorted players who made up the fruit trade. Once unloaded, the boxes commanded little attention, their colorful labels desperate to lure an interested eye amid the furious activity of the Terminal Market. This was the dwelling place of receiving merchants—the people who took charge of the fruit but not possession of it.[18] These handlers (also referred to as brokers and commission merchants) located the product on the wharf and arranged its sale for a percentage of the final price. Bridging the commercial distance between west and east meant passing through gates monitored by these middlemen—the only people in the economy who knew both supply and demand, producer and consumer. There is no way to understand the marketing crisis and the subsequent movement to organize fruit growers into crop-specific cooperative corporations without understanding their role.

The payment of a commission evolved along with competition among growers. In the earliest years of the trade, produce like grapes sold at private sale to the affluent who ate them only occasionally and primarily for dessert.[19] Merchants purchased directly from growers because they could be sure of finding buyers and because the potential profits justified the cost and the risk. But as the number of producers increased and prices fell, as markets became more competitive and profits less certain, merchants became less willing to absorb the losses when shipments spoiled. In other words, merchants stopped risking their own money when they knew that growers would bear these costs plus a commission in order to dispose of their crops. As more and more shippers appeared and sought the attention of fewer consignors, the advantage turned to the buyer.[20] Merchants no longer paid for fruit outright because they no longer had to.

This turnabout put merchants on top of the trade and exposed their many deficiencies. Back when payment for their crops came "cash on delivery," few growers bothered to notice that fruit was not the kind of product commission merchants traditionally sold. It had to be turned over quickly; it required special care and storage; it needed to be represented to increase its distribution. Commission merchants did none of these things and took no responsibility for the outcome. Shippers complained that they paid consignors even after a total loss.[21] The wholesaler—the man in the middle—emerged as the pivot in managing the trade, and growers developed a profound distrust for almost everything merchants did and said. As long as they stood in the critical space between California and urban consumers, "the industry was at its lowest ebb."[22]

To farmers all over the country, the commission merchant represented the perniciousness of illegitimate profit. Growers shared in this opinion and considered graft among consignors an open secret. One editorialist spoke for many others: "They can write you sweet letters of undying devotion, but if raisins are going up, they unfortunately sold the most of yours (they don't yet say how many) at the low price. If raisins are going down, none has been sold until hardpan is reached, when they all go. I would like to hear from the shipper that ever got anything above the lowest price that ruled in the market for his raisins during the time they were 'on sale' in brokers' hands."[23] The list of supposed crimes committed by commission merchants is long indeed. They lied about the condition of fruit on arrival, selling a perfectly sound carload on their own account and remitting pennies or nothing at all; they failed to store fruit properly, ruining it after a successful journey to the market city; and they took rebates from shippers who wanted special treatment.[24] Consignors did not keep accounts of prices, amounts, or dates of sale to establish the legitimacy of their dealings. When the news over the telegraph read: "The raisins offered at auction on Tuesday did not find a bidder," the shipper had to depend on the integrity of the merchant with no way to verify the actual circumstance of sale.[25] No government agency investigated claims of fraud, and no other method existed to bring fresh fruit to consumers in the most concentrated markets in North America: "Like it or lump it," as one shipper put it; "no satisfaction and no redress."[26]

Selling agents, no matter how corrupt they may have been, struggled to make a living in a system they could not have manipulated had they wanted to. The differences in price one day to the next often had to do with unanticipated glut or dearth. Any major city could absorb five carloads of peaches, but put fifteen of them there at the same time, and not one of them would pay off.[27] When fifteen cars rolled into New York at peak season they met produce from other regions and clogged up the city. The only people who benefited were the peddlers and street venders who bought the best of California at a lower price than it could be purchased in San Francisco. The needs of any given city on any given day for any given commodity could not be consistently predicted.

Here is a case study in market mayhem. The New York merchant house of Sgobel and Day took matters into hand and installed Herbert Fairbank as their agent in Sacramento to collect telegraph reports of conditions in cities across North America. Fairbank passed the messages on to clients like R. B. Blowers and Sons, the fruit shipping company in the Sacramento Valley. Rather than offer certainty to the shipper, Fairbank's quotations revealed the volatility of supply and demand. For example, in September of 1892 Bartlett pears sold by the box for between $1.30 and $2.28 in Chicago, but in New Orleans the market "completely collapsed" because of oversupply. Pears for sale in Minneapolis fetched no higher than $1.75, but just across the river in St. Paul they sold for $2.60. And this information only told shippers what had happened *the day before*. They still had no idea how many cars from producing regions across the country were on the way. By the time a packinghouse responded to a query like "Cannot you get us some fruit? Market practically bare," its cars might have ended up last in line to be unloaded and last to be sold. This was not merchandising, wrote C. C. Teague, but gambling.[28] Commission merchants alone could not be blamed for such chaos, although they often contributed to it.

Harris Weinstock searched for a way to circumvent the power of commission merchants. Convinced that the system of private sales would soon destroy the trade, he advocated the auction. Citrus from Spain and Florida sold at auction and brought higher prices than the same from California. Weinstock told readers at home that the best produce in New York went at auction and that "prices at private sales would not, by 20 per cent, net as much as they do at auction."[29] Wherever it went, California fruit competed against itself. Buyers went from merchant to merchant comparing quality and prices and bargaining down everyone on the wharf. In effect, the auction helped producers by shifting the burden of competition from seller to buyer. With buyers all in the same room bidding against each other, prices reflected competitive demand as well as competitive supply.[30] The auction brought higher prices, but it still left the consignment system intact, and it only passed the product to the next set of hands.

Those hands belonged to the jobber. Jobbers purchased the fruit on their own account and sold it to retailers who, in turn, sold it to the public. They occupied the space between the terminal market and the neighborhood wholesalers, where grocers and restaurants came to assemble their stock. By the time a jobber purchased ten or twenty boxes it may have been three weeks since the fruit left California, and those apricots looked better in the auction room than they did the next day on the floor of the jobber's store. In time of glut, boxes sat piled high in the selling lot, and jobbers had them stacked out their doors and into the streets. The jobber conducted a fast turnaround, high-volume, low-margin business. He knew the neighborhood stores and furnished what they asked for.

The only California fruit most retailers had ever seen in the 1880s was expensive and soft. Grocers treated it more like caviar than a common element of diet. In their minds, grapes and oranges that had been "imported" commanded a clientele all their own. When Harris Weinstock visited a street-side stand on Broadway he learned that the grocer had no interest in buying more produce from the Pacific Slope—even for a lower price—because he did not think it would sell. Weinstock suggested that grapes might one day sell as low as fifteen cents a pound, but the grocer still would not buy much. At a price so low grapes would "lose caste among the finer trade who buy it now at a fancy price because it is kept only in first-class places." Weinstock heard the same story all over town. Retailers liked California fruits expensive and in small quantities rather than cheap and plentiful, and they made no attempt to promote them.[31]

As dismal as the market saga seemed, it got even worse. Soon after leaving Chicago, on his way to New York, Weinstock toured Michigan's fruit districts and found—to his horror—that its growers produced peaches of high quality. Michigan was supposed to be one of those pitiable little places where tragically misinformed people attempted horticulture amid snowdrifts. Weinstock was stunned: "Before entering Michigan the writer was filled with the traditional California conceit, and honestly believed in common with thousands of fellow-Californians that there is no other spot in the land so highly favored for fruit, and especially peach culture, as California." He saw fifty thousand acres on Lake Michigan, all of it "peculiarly adapted to the culture of the peach, quince, plum, pear and vine." The area had the potential to yield 25 million ten-pound baskets of peaches—"enough to supply the whole country." Other states also supplied peaches: Ohio, Delaware, New York, Indiana, Kentucky, Missouri, Kansas, and all of the Southern states and Maryland. Georgia had not yet become a major producer, but it was on the way.[32]

After two days in Michigan Weinstock had to confess that "providence did not intend giving California exclusive control of all fruits, but that our Eastern brethren have been permitted to share some of our privileges."[33] It was an important admission. Michigan had an advantage over certain varieties of peaches and grapes, and Weinstock called it pure folly for Californians to grow them. Varieties like the Black Hamburg, Black St. Peter, Black Pinot, and Malvoise could only be sent east of the Rockies "at a loss to grower or shipper." Rather than invite a competition between natural advantages that California might lose, Weinstock warned his readers back home to stay close to their specialties. As Oliver Baker might have said, the direct communication between regions had served to enforce their agricultural differences.

All the troubles he had seen between Chicago and New York convinced

Harris Weinstock that a reversal of fortune required much more than an auction system and the Fruit Union. As he cultivated the indignation of his public, the emissary also told them that their products had to change in order to survive in the cruel world of commerce. He began to advocate a different conception of quality: "The grower who thinks he can push anything he pleases on the Eastern market and get a profitable return makes a sad mistake." Only by putting "brains and conscience" into their fruit could western growers hope to prosper. By this he meant uniform grades to define quality, clean and standard boxes, orderly promotion, and all the other things that helped to rationalize business and eliminate waste. People in the business needed to alter the product to the market as badly as they needed to alter the market to the product. In short, competition with other favorable climates demanded that California growers expand the kinds of activities they were willing to engage in order to sell fruit.[34]

The failures of the present system could not have been more obvious. A survey of 365 railroad cars loaded with California fruit conducted during eight days in July of 1902 found an average net loss per car of $142.55.[35] The carloads all left during the same week and sold in eleven cities. In Philadelphia, Cincinnati, and Baltimore, not a single carload made money, an astonishing loss considering that the survey was conducted at peak season—the time of year to expect returns that would balance the costs incurred during the growing season. With no way to coordinate producers and shippers to avoid flooding and with no organized campaign to stimulate demand through advertising, growers recorded losses in spite of all the advantages of their location, in spite of the quality of their produce. Eight days in July could make or break a business. If a deciduous orchard was not making money in July, it might not be making money at all.

The depression of 1893 compounded every problem. Between 1888 and 1893 raisins sold for about $.05 a pound on the vine; after 1893 there came a swift decline until the price had fallen to ¾ of one cent per pound in 1897.[36] The glut meant ruin all over the state. Seeking to ease their pecuniary strain with lower rates, the members of the Fruit Growers and Shippers Association complained to the railroads that the facilities for marketing deciduous fruit had all but broken down: "*losses to the growers who shipped have been the rule, and profits the exception.*"[37] All of which made raisins so cheap in 1900 that growers sold them as feed for hogs and horses. Thousands faced bankruptcy. Hundreds of mortgages foreclosed, entire vineyards were uprooted, and the Fresno County Assessor's office reported a reduction in raisin grape acreage of sixteen thousand acres over two years. Desperate growers and their families left rural counties or turned adrift to compete for jobs with Chinese, Japanese, and other immigrants in the state's migrant labor market, "tramping about the State in search of a day's work." [38]

No region of specialized agriculture could survive these conditions for long. [39] Increasingly, growers, shippers, legislators, bankers, and chambers of commerce began to talk about the perils of distance and the commission system and how they could be overcome. At stake was one of the state's most promising industries, hundreds of millions of dollars in real estate, and the entire economies of districts firmly invested in fruit growing, such as the uplands of Southern California, the San Joaquin Valley, and the orchards along the Sacramento River. By the late 1890s business leaders and boosters began to assert in public that they could see only one way for producers to place California agriculture on a stable basis: take control of distribution.

In 1897 growers met in Sacramento to hear J. A. Filcher, manager of the state Board of Trade, speak about agricultural marketing. What he told them defined the goals of a new project. "Our efforts thus far in California have been largely devoted to considering what to plant, how to plant, how to cultivate, to irrigate, to fumigate, to prune, to pick, to cure and to pack." Filcher spoke of fresh concerns, like "extending our markets, and . . . increasing the demand for our products" and of cooperation as a commercial tool. [40] Cooperation called for the federation of many producers to attain a measure of authority over the price of fruit and the condition of its sale. If growers could combine their numbers, said Filcher, "the day would be much nearer when this State could dictate the prices of canned and dried fruit to the world."

Put in other terms, those who raised specialized crops needed to think past production. They needed to practice the intensive cultivation of desire on the consuming end. Just like the drive to take fatter harvests from limited land, advertising aimed at increasing the demand for fruit among a limited (though still growing) population. Filcher argued that the effort to gain control of disposal called for new capital and energy and required two linked activities. As he put it: "Let me suggest to the fruit growers a motto or a guide to action . . . 'organize and advertise.' " [41]

Just a year after the fall of the People's Party and the election of William McKinley and Theodore Roosevelt, a different kind of farmers' movement took hold in the Far West. Rather than summon lofty invective against "colossal fortunes" whose possessors "despise the republic and endanger liberty," as did Ignatius Donnelly in 1892, the business booster summoned self-interest in language that any Gilded Age magnate would have cheered.[42] As a rural movement, cooperation would bear the mark of California's social conditions and would reflect the complaints of orchard capitalists. By the beginning of World War I, the cooperative association had proven itself in selected commodities and began to change the relationship between California growers and the markets they so badly wanted to reach. Its creation was also a step in the direction of a more highly managed countryside.

Cooperative Corporations

With all the fervor of rural reformers, California fruit growers united to make the American economy a safe place for specialized crops. They proceeded with the blessing of the Country-Life Movement whose motivating spirit, Theodore Roosevelt, exhorted American farmers that their future prosperity lay in forming an effective counterbalance to the power of the industrial trusts. Careful to express the proper deference for "rugged self-reliance," Roosevelt offered that the farmer "must learn to work in the heartiest cooperation with his fellows, exactly as the business man has learned to work."[43] Others also proposed that farmers acting in combination might derive some of the advantages of big business while maintaining themselves as independent producers. Kenyon L. Butterfield, like Bailey a professor of agriculture and an advocate of progressive rural life, argued that "the farming class must produce as a unit" and encouraged the formation of a coordinated agricultural industry that would protect farmers against big business.[44]

Roosevelt and Butterfield took up the trope of cooperation. For them it described any farm-based organization that made no claims on Wall Street, railroads, urban consumers, or government. "Cooperation" sounded like good old-fashioned pulling together to help out neighbors during hard times. To others it signified a belated attempt by farmers to join industrial capitalism. Edwin G. Nourse wrote extensively on the subject of cooperatives, often asserting that they presented decentralized farmers with a scheme of organization that enabled them to participate in the economy as equals along side colossal, managerial firms.[45] Through Nourse's lens, cooperation delivered a prosperity to farmers that populism had only promised, and it left the politics out. In this way, it seemed to represent a final coming to terms with the world as it is and the end of agriculture's prolonged adolescence.[46] That Roosevelt, Butterfield, and Nourse mostly agreed about the apolitical ends of cooperation suggests its appeal to progressives. Though the various supporters of the Country-Life Movement never approached the subject with one mind, many recognized cooperation as a blending of rural organization with business pragmatism—a conservative response to populism.[47]

Liberty Hyde Bailey also made comment on cooperation, but while other country-life leaders feared populist politics, Bailey feared industrial capitalism. Though certainly no populist, he seemed more comfortable with the idea of a farmer's party than he did with the idea of farmer corporations. Bailey's brand of rural romanticism brought him to recommend cooperation only as a tool to unify economy and society in a fractured countryside: "The essential thing is that country life be organized: if the organization is cooperative, the results—at least theoretically—should be the best; but in one place, the most needed cooperation may be social, in another place educational, in another religious, in another political."[48] The cooperative, in Bailey's corrected form, brought farmers together to make agriculture more pleasant and less tedious; it facilitated social contact and helped to spread practical knowledge; and it functioned best when it resembled the local grange. What this hesitant endorsement does not allow is for the cooperative to act as an agent for farmers in the marketplace. Such an organization might help to solve certain commercial problems, Bailey believed, but a cooperative was "properly a society, rather than a company." Any federation of farmers that produced and sold as a corporation could be considered cooperative "only in name."[49]

None of this mattered very much to orchard owners in California who combined their numbers and pooled their crops for the sole purpose of improving the outcome of sale. The form they adopted can be traced to Rochdale, England, where in 1844 a group of English textile workers went into business for themselves to secure their independence from low wages and tyrannical employers.[50] With only slight modification, the same "Rochdale Fundamentals" that founded the first such association founded the marketing cooperatives that appeared in California beginning the 1890s.

First a definition. True cooperatives return all profits to members after deducting operating expenses. Cooperatives distribute earnings pro rata, or according to the amount of business each member transacts through the association, and keep nothing for themselves. Only those who produced the commodity that the cooperative is established to distribute can become members and stockholders; no one outside of the association can own its stock. All members hold the right to vote, but some cooperatives set voting power proportional to the number of acres or the amount of business that each of its members contribute, suggesting that more affluent growers would have more influence.[51] A cooperative contracts with its members for a specified amount of fruit to be delivered to packinghouses and shipping points. The central organization then advertises, bargains for prices, searches for new outlets, arranges for sales, and remits profits.

Put in other terms, a cooperative is a nonprofit collective. The marketing association serves as a commission house that takes only enough commission to maintain its services. In the words of G. Harold Powell, general manager of the California Fruit Growers' Exchange, "a co-operative organization . . . is not a corporation in which the capital is contributed primarily in order that it may earn a profit . . . nor one in which the producer's product is handled by a corporation for the benefit of the stockholders rather than for that of the members."[52] Typically, they owned no capital stock, issued no tradable shares, and thus never acted in the interests of anyone other than their grower-owner-members. Control vested in the membership maintained the independence of the association from shippers, canners, speculators and others with diverging interests.[53] Fruit growers in California applied these principles to the necessities of their own time and place and founded a series of companies that the Rochdale textile workers would have barely recognized.

More than any other person, G. Harold Powell defined the philosophy and the goals of cooperation in California. Born in New York, Powell graduated from Cornell University in 1895 where he studied horticulture with Liberty Hyde Bailey. A master's degree followed and then a position at the newly created Bureau of Plant Industry at the United States Department of Agriculture. Powell first became interested in the California orange industry while studying cold storage in 1904, and he made a visit to Riverside that year to advise growers on refrigeration. He moved his family to Pasadena in 1911 and took over the management of the California Fruit Growers' Exchange in 1912.[54]

Powell and the orange growers gave flesh to the conception of cooperation as a collective business enterprise without (explicit) political aims. In a word, he argued that any successful federation of farmers should be business-minded and highly specialized. Powell wanted no one to confuse his industrial capitalists with the advocates of free silver and the subtreasury.[55] He believed that cooperatives should be founded for tangible economic reasons, not out of principled outrage. "American agriculture is strewn with the wrecks of associations that were the outcome of high motives and impractical enthusiasm," thumped G. Harold Powell.[56] And he called the Farmers' Alliance and the People's Party evolutionary dead ends in American agriculture's politically charged history. Individualist growers did not unite to beat the trusts but to join them as a means of survival in a world where only combined capital survived. Sounding like the economist Nourse, Powell insisted that "capital has been concentrated around gigantic undertakings," and individuals stood not a chance in the shadow of Leviathan enterprise.[57] The tiller reborn would be a stockholder-producer in his own rural corporation, a new identity in changing times to bring him closer to integration into the nation's industrial sector. "In place of transacting business man to man as his father did before him," Powell wrote, the farmer "has become a more or less important part of the scheme of modern industrialism. He is no longer isolated. He is a link in the modern industrial and social chain."[58]

Powell called "common purpose" the first foundation of cooperation, by which he meant a shared experience with the many difficulties of horticulture and a shared desire for pragmatic solutions. The fruit union had been an experiment in common purpose. Its failure convinced growers that general marketing made about as much sense as general farming. No single agency could master the complexities of selling products as diverse as almonds, lemons, and plums. Rather than organize around common problems, said Powell, growers needed to organize around *specific commodities*. Powell asserted regional specialization as the key to successful cooperation: "It is only when they become specialists in a crop . . . and have to develop special facilities for the handling and distribution of the crop, that a group of farmers have a common purpose comparable to the aims of a large manufacturer."[59] Any lasting agricultural industry "must be founded on a special industry," such as oranges, pears, or raisins.

The hard-nosed demeanor of the California cooperatives prompted economist Ira B. Cross of Stanford University to point out that "even among the cooperators themselves a surprising lack of the true spirit of mutual helpfulness is evident." Cooperative enterprises in California, wrote Cross, "seem to have been organized with but one object in mind, and that is to make money for their stockholders. They are mere profit-making associations."[60] Powell would have agreed. Suspicious of ideological fervor, he stripped the cooperative of high motives. Under his leadership and with urban, sophisticated growers as owners, a vehicle for rural uplift became a single-minded business venture.

Progressive government in California quickly recognized the power of pragmatic cooperation and moved to support it. Where Powell gave the marketing association its philosophy, Harris Weinstock gave it the assistance of the state. In 1916 Weinstock became the first state market director. The appointment came from Governor Hiram Johnson, who surprised no one by naming a dry-goods merchant to pursue the state's interest in agriculture. Weinstock recognized that the benefits of cooperation had not spread far beyond oranges and raisins. Through his bureau, the state of California expressed its interest by extending help to new associations at the moment of their formation.

The producers of peaches, prunes, apricots, and pears still sold fruit the old way, without exerting influence over commission merchants and without promotion. Weinstock's office, the administrative arm of the State Commission Marketing Act of 1915, brought government into the business of agricultural marketing. After wrestling with the legislature for a year to redefine the goals of the bureau as well as its name, and with the advice of the chief council Aaron Sapiro, who served as the attorney for many California cooperatives, Weinstock found himself in a position to bring order to the countryside in a way that no country-life reformer could have imagined. As head of nation's first state market commission—charged to advise, promote, assist, foster, and encourage "the organization and operation of cooperative and other associations"—Weinstock's purpose was nothing less than to change the relationship between California and the American economy and to build agriculture on the West Coast into a major industry. He acted on the premise that what minimized "the waste and expense of distribution, [worked] to the benefit of the consumers." Along with Powell, the market director defended the cooperative as a tool in the public interest and envisioned a collection of regionally based agricultural companies geared for the steady distribution of the American food supply.[61]

Yet the rise of cooperation had a more subtle effect. Like any other corporation, the cooperative set out to standardize the quality of its product and exert greater control over every stage of the process and promotion. In this way, the cooperative confirmed the single crop, making it the very preamble of its charter. The American food supply and the California landscape reflected its efficiency and simplicity, a necessary transformation, perhaps, in the conquest of distance.

A Deal in Pears

Cooperation translated soil regions into corporate geography and institutionalized natural advantages. No longer simply a local pattern in trees and vines, specialization became the basis of an industry. Single-crop marketing associations changed the commercial prospects of California fruit, and even their very nature. Such a vast story needs an example to find its focus, and the case of the California pear growers serves this purpose. A single deal in pears connects Weinstock's career in state government with the formation of a successful cooperative and the onset of the postwar farm depression. We are searching for the relationship between mass marketing and the landscape of rural California, and pears can help us to make the connection.

In 1916, Frank T. Swett, a pear grower and horticultural commissioner in Martinez, Contra Costa County, approached Weinstock for help in organizing a cooperative. Pears lacked the distinction of oranges and raisins among consumers. Though attractive and good to eat, they perished easily and thus came to be sold in cans. Whereas oranges became famous for their fresh-as-sunshine color and raisins for being cured in the California sun, pears from the Pacific Coast entered the American diet as a processed food: peeled, sealed, and cooked in tin. The canners took tender pears through their stainless-steel dynamo and gave them eternal life. They sold a commodity that could be stored for better prices and traded on an exchange. Pears in cans could travel to places where fresh pears had never been seen. And though producers did not benefit from the higher profits of final sale, neither did they worry about finding buyers or collecting returns. Canners took their cut of the profit just like commission merchants. When they enjoyed the commercial advantage, the anguish of the distant market felt close to home.

Swett reported to Weinstock that the pear industry "was in a chaotic and disheartened condition, with the unorganized growers practically at the mercy of the canners."[62] The problem was an old one for farmers. Organized and capital-rich buyers manipulated poor and isolated producers. Pear growers, however, already had the seed of an organization to protect their interests—they often sold as members of small packing associations. Companies like the Contra Costa Fruit Growers Association, the Empire Vineyard and Orchard Company, and the Elk Grove Fruit Exchange formed in the 1890s, but each consisted of little more than a lithographed label and a packing shed.[63] These outfits never found the stability they sought. Nonetheless, Frank Swett knew that local associations might one day form the commercial backbone of a pear cooperative. For a model he only needed to consult the orange growers. They used neighborhood packinghouses as the local units of their centralized fruit exchange. Members living in orange districts from San Juan Capistrano to Tulare sold fruit under their own brand names and waited for shipping instructions from larger district exchanges. The corporate offices coordinated this federal system, ordering carloads to cities according to the best market information available and advertising the product to promote its many uses.[64] Swett wanted to do the same with pears, but few producers would follow him.

Over the next two years a series of ambitious mergers changed the balance of power in the fruit industry and forced pear growers to federate. The California Fruit Canners Association, a collection of eighteen small companies, combined with Central California Canneries to form the California Packing Corporation in 1916. Under the ownership of George Armsby, whose merchant house served as the wholesale department of the new conglomerate, and with $16 million in guaranteed loans from Wall Street banks, "Calpak" emerged as the largest canning company in the world. The firm had sixty-one plants, located all over the West, including fifty-three in California and one each in Hawaii and Alaska.[65] Its Del Monte brand brought nationwide recognition and $6 million in net earnings in 1918.[66] Calpak dominated an industry that also included the California Cooperative Canneries, Caladero Products, and Libby McNeill Libby—each of which purchased pears.

As long as they liked the prices that the canneries dictated, pear producers had little incentive to cooperate. In 1916, at the time of Calpak's formation, canners big and small competed with each other to purchase the best pears of the season, and the average price they paid that year reflected a seller's market: $50 a ton for Bartletts. That was money enough to satisfy most growers that they needed no organization. But when it came time to figure a price the following year, the canners came together. Led by Calpak, the major firms offered the growers an average of $35 a ton—$15 less than the year before. A second year at the same level finally gave Swett an opportunity.

On 2 April 1918 the directors of the California Pear Growers' Association met in Harris Weinstock's San Francisco office to write bylaws and develop a plan to sign up five thousand acres, the minimum number required by law to become a recognized cooperative. For two months they hired solicitors to go orchard to orchard; they offered membership to Chinese and Japanese growers; and they looked to Kern, Los Angeles, and Inyo Counties for wayward members.[67] Orchardists joined with the assurance that they could market their pears fresh in any manner they chose and that contracts with the association for canned pears bound them only year to year. They hesitated, in short, before signing over their crops to an organization with no record.

In June the board of directors, having satisfied the acre requirement, held their first meeting in rented offices and made the kind of decision they were founded to make—they set the price of pears: "Resolved that [the] price for standard pears [measuring] 2¼ inches and up [be] fixed by the Pear Growers' Association for the season of 1918 [at] $70 a ton."[68] Once a price had been agreed upon, the association went to its members to assemble the shipment. The canneries tried to break the association with "false and misleading propaganda" intended to dissuade orchardists from joining.[69] As long as the federation continued to deliver higher prices than most growers could secure on the outside, the cooperative held.

The pear growers found themselves in an admirable position.[70] Prices for 1919 hit $116 for the best of the crop, and some expected the 1920 price to top off at an astounding $150. The volume of the business had grown very quickly, riding the war years and new European markets. In 1920 the associated pear growers grossed $1.2 million.[71] They had signed up about one thousand growers representing close to twenty thousand acres. Even more important, they had established themselves as a successful bargaining agency that negotiated prices and amounts with some of the largest canneries in the country.[72] In 1919 California produced 30 percent of all the pears grown in the United States, more than any other single state, more than New York, Washington, and Oregon combined, and demand seemed to have no limit.[73] The market had been reached and mastered, or so it seemed.

From Overproduction to Underconsumption

The deal in pears brought the people who grew them only so far in capturing influence over the market. As Harris Weinstock discovered between Chicago and New York, the problem of the market was really two: disposal and volume. Cooperation did not change the fact that the single-crop farmer depended on a greater economy, one beyond the influence of any single corporation. If growers wanted to survive as manufacturers in a world of manufacturers, they would have to influence the consumption of their products in the manner of manufacturers. The question of volume and advertising became urgent at the time of the Farm Crisis and the rural depression that followed the end of World War I. However, where the price panic motivated farmers in other places to demand political reparations for a dysfunctional economy, California growers asserted their faith in the compatibility of agriculture with industrial capitalism. They looked at the problem as one of low market share, and their response is also part of the industrialization of agriculture.

Just as a workable system of pooling and negotiation emerged, the market failed. The problem had to do not with broken channels of commerce, but with the demand for fruit itself. During the war, orchardists enjoyed the same rise in prices as other farmers, and like other farmers they planted with the expectation that good times had come to stay. The pack of canned fruits and vegetables from California doubled from ten million cases in 1915–16 to 20 million cases in 1919–20, but when the war ended the trend ran out of steam.[74] At first growers assumed that slow fresh sales could be offset by sending the surplus to canners.[75] But the canners felt the same slump. Only the year before, canners had placed orders for more pears than the land could provide, but in 1921 the market for canned fruit busted. Foreign producers went back into production, the army no longer needed supplies, and Americans found themselves deluged with fruit cocktail in syrup. When demand fell apart it caught canners with contracts to buy crops that they knew they could not sell. Half the pack of 1920 still sat unsold in 1921, and Frank Swett said that unless some of that fruit was "eaten up" before the next harvest, growers would have to find other ways to sell it or suffer calamitous losses.[76]

Trouble in the fruit industry appeared in the pages of the *Western Canner and Packer*: "Canners Need Financial Assistance," read one headline.[77] They assisted themselves by keeping wholesale and retail prices high to cover their costs. Cans, energy, fruit, labor, paper, sugar—all made grocery store prices higher than consumers were willing to pay. In May of 1914 a can of California "extra standard" No. 2½ pears cost $2.35; in May of 1920 the same item cost $5.10. Cling peaches went from $1.90 to $4.00 during the same period, while the trade complained that nothing sold, that shoppers stayed away, that warehouses were filled with leftover stock from last month and last year.[78] Frank Swett estimated that Dole lost $10 million to $12 million and that Libby McNeill Libby lost

$10 million in 1920.[79] Promoters of the canning business in California refused to believe their figures and still talked about a "resumption of cannery building on a large scale," claiming that the prevailing conditions had a "false basis."[80] There was no false basis. The crash brought into question, yet again, the feasibility of specialized agriculture, and it forced pear growers to find their way without the canners. The association had established itself as a successful bargaining unit, but to weather the bust its members had to take greater responsibility for the eastern sales of their fresh fruit.

For once growers could not blame bad times on commission merchants. By the 1920s the commission business had fallen into decline. With USDA-enforced grades and standards, merchants could no longer practice wharf-side evaluation, could no longer reject shipments for their own reasons by falsely claiming poor quality. Other connections also weakened the old reign of the consignors. Shippers made brokerage connections in distant cities; buyers visited California where they installed agents; market news services provided accurate information. All of this made players in the trade more willing to purchase fruit directly from producers based on carload prices quoted over the wires.[81] Even the capricious rails became more predictable. Refrigeration improved when the transport companies and the growers worked together to guarantee speedy delivery.[82]

In short, the commission merchants watched as the world they had once dominated disappeared. In their own defense, the consignors compiled facts and figures to justify their role, claiming that the true bastions of illegitimate profit were the railroads.[83] They finally appealed to G. Harold Powell, who called eliminating the wholesale merchant "impractical." "We specialize in shipping oranges and lemons," he explained; "we cannot go to retailers or customers direct, and no other organization that specializes in one or two products can do it and succeed." In the end, growers did not have to destroy the merchants, they only had to reform them. The merchant house Sgobel and Day survived by reinventing itself as a "fruit distributor" in 1915, a title better suited to the new order of things.[84]

Although corruption and inefficiency no longer fouled the waters of commerce, another problem, one endemic to agriculture, forced new ways of thinking about fruit and new ways of selling it. Fruit growers could not regulate their production without tearing out trees and vines in a destructive measure to balance market forces.[85] They remembered the depression of 1893 when a financial panic at a time of rapid expansion made the fantastic yields that once added to California's stature a terrible burden. Thousands of acres planted in the booming 1880s came into bearing just as the depression hit. The price of a box of pears fell from $3.10 in September of 1892 to as low as $1.40 in October of 1894.[86] Half the pear acreage had to be dug out; half the prune and plum trees in the state had to be burned or cut down.[87] No blight took these trees. The preservation of the industry demanded that they vanish.

Following the boom of the Great War came a similar crisis. Fabulous profits attracted people to California to buy into prosperity. Seeing a chance to revive the furious speculation of the 1880s, local business spoke the old language of advantage and progress. Said one investor in real estate during the first the days of the unfolding collapse, "The eastern tourists are familiar with our climate . . . , there will be millions of dollars invested in orchard properties, and it is to the interest of every loyal Californian to Boost for the Cause."[88] Between 1910 and 1920 thousands of acres of fruit trees went into production. But boosting the cause became a futile business after 1921. By then, growers again had to contemplate whether or not to tear out their trees.

In 1921 the orchard capitalists decided to go about righting the balance by other, more constructive methods. They declared that the problem was not one of overproduction but underconsumption, and they chose to invest in consumer "education," by which they meant advertising. In 1900 few Americans had eaten fresh pears from California. In 1920 they consumed them, if not daily, then at least from time to time. The managers at the pear cooperative figured that per capita consumption averaged one-fifth of a can per year and one pound fresh, a rate far too low to absorb the fast-piling pears. Four thousand carloads of Bartletts traveled to points east in 1920, and if the canners did not buy up the crop of 1921, eight thousand carloads would cover the same territory.[89] To make room for the surge in produce soon to arrive in the East and to compensate for the increase in production that always followed good times, pear growers did what the producers of other fruit crops had done long before—they created an image for their product and put it before the public. In addition to reorganizing the structure of the trade, cooperatives cultivated hunger on the consuming end.

Fruit became an object of consumption. Bright and colorful, sweet and compact, versatile and healthful, fruit needed little consumer education to sell it. Before cooperation, orchardists dumped their produce at the feet of receivers and jobbers who treated it not much differently from iron rails or bales of cotton—durable commodities the demand for which could not be increased through tactful persuasion. The grocer's display no longer accomplished the goals of a mass-producing countryside. Families in the city chose their diet from foods grown near and far, eating in one meal what required the myriad lands of the continent to produce. This meant that they had no reason to eat any one thing in particular. With milk and eggs, one's choices remained limited, but with fruits and vegetables, whim and impulse ruled the stomach. Fruits that once announced the peak of summer in neighborhoods and shaded valleys all over the country, whose ripeness helped mark the passage of time for the people who grew them, now competed with imports from other climates and soils for the attention of the multitudes passing on the street.

One of the most important functions of marketing cooperatives was to promote fruit as items of pleasurable consumption. The continuous expansion of the fruit industry called for growers to appeal directly to the tastes and fancies of people who had never thought much about pears. For unlike other products of the farm that got processed, mixed up with other commodities, and resold as mayonnaise, bread, or soup, fruit off the tree could be sold as a finished product with a brand name, a wrapper, and an attitude. The old marketing offered fruit to a hungry public; the new marketing encouraged consumption for its own sake and employed publicity, name recognition, uniformity of product—merchandising.

Advertising began with labels. Where subjective judgments prevailed, image often made the difference between profit and loss. In a business where products of widely differing quality competed, brands helped wholesalers to distinguish the shipment of one grower from that of another. Professional lithographers designed these miniature billboards to create a pleasant impression in the auction room or on the floor of the produce market.[90] The art of label advertising reached its perfection in the person of Don Francisco, at the age of twenty-four the first advertising manager of the California Fruit Growers' Exchange and later the vice-president of the firm Lord and Thomas in San Francisco. Francisco published a booklet in 1918 called *Labels—Suggestions for the Shipper Who Is Seeking to Give His Pack a Worthy and Effective Mark of Identification*.[91] He argued that labels should be designed for the jobber and not the consumer. Shippers often liked to feature mountains, orchards, and other landmarks evocative of their homes, but Francisco recommended otherwise. Labels should attract the eye from ten feet away when stacked five boxes high and ten boxes wide.[92] Subjects should be simple and distinctive, with no detailed images. Orange labels that conformed to Francisco's principles often featured the fruit over a dark background to highlight its brilliant color and attractive texture, along with another subject like a pointing hand. Brand names like Order, Have One, and Demand were typical of the 1920s.[93]

Yet the image on the box did not reach the public. To do that, cooperatives invested in full-scale national advertising campaigns, waged in newspapers and magazines, and designed to increase consumption. Frank Swett argued before the membership that they could expect to double the consumption of pears within a few years through direct advertising.[94] He relayed the success of the strawberry growers who bought banners and newspapers to promote themselves locally: "One day's slump in the market would have cost the grower more than the cost of the whole season's advertising. . . . It paid big dividends."[95] The association opened an Advertising Trust Fund with $53,492 raised by assessing packers and canners $.05 a box and 5 percent on canning deliveries. (Under the direction of Don Francisco, the orange growers raised $400,000 for their campaign.)[96] They would use the money to bring news of California pears to people who knew little about them, to show how beautiful the fruit looked on a table or plate, and to suggest how people might eat them week to week. The fund represented a good start, noted Swett, but a prolonged increase in consumption implied an extended campaign.[97]

They started with Philadelphia and Boston. Both cities comprised sufficient urban and suburban populations to justify the expense—Philadelphia counted seventeen hundred grocery or fruit stores and Boston included over two thousand.[98] But pear growers faced the same attitudes that Harris Weinstock found on his trip to New York in 1886. Retailers still hesitated before handling "any fruit that is more perishable than an apple," and "to reach these men," wrote Swett, "is a big task."[99] In January of 1921, the association initiated a correspondence with storekeepers in the two cities and placed its own representatives on the streets of Philadelphia. The cooperative offered merchants a set of "store helps," consisting of posters and signs, to draw the attention of shoppers.[100] The Schmidt Lithograph Company distributed advertisements to grocers and placed ads in newspapers. Printed media provided the campaign's second front. Swett hired "advertising experts" who designed the pitch to "catch the eye of the hasty newspaper reader." Concern over how the pears would be promoted settled on their freshness, their refreshing qualities, and their low cost.[101]

The new marketing stressed "education" over selling. Direct advertising told people why to eat fruit, how to eat it, and even with whom to eat it. Companies like the peach growers initiated special weeklong public-relations invasions in target cities to acculturate consumers. Still, the enthusiasm of the public had its limits: "It hasn't been the heavy receipts that kept down prices," commented one handler after Chicago survived Peach Week, "but it is obvious to me that the people have been peached to death."[102] Nonetheless, such campaigns worked to the advantage of merchant and cooperative. When the California Associated Raisin Company sent a recipe for raisin bread to bakers all over the country, the bakers used it as a way to sell a more expensive item. The bread had a spillover effect for Sun-Maid's boxed raisins, "all of which merely proves," wrote the *Packer*, "that in modern business there is no force as powerful as advertising."[103]

The pear campaign blitzed still other cities as it moved into 1924, and it began to show results. Not only were other major cities added to the list, "but also adjoining cities in the trading radius, and even beyond." Twenty-one young men working under a local superintendent covered Chicago's Elevated Railroad platforms with large lithograph posters "which attracted the attention of hundreds of thousands of commuters daily," nine men went to Boston, seven canvassed Philadelphia, and ten brought news of pears to Cleveland and its suburbs to a radius of fifty miles, including Akron. Swett told members that the target cities purchased more than twice as many pears in 1923 as they had in 1920—a 119 percent increase. He compared this to the 46 percent increase in New York where the association spent no money on promotion.[104]

In the ultimate marriage of agriculture with the cutting edge of consumerism, the California Pear Growers Association hired the J. Walter Thompson Company of San Francisco in 1927 to conduct market research. J. Walter Thompson discovered that people ate their pears "as is" (89.1 percent), in salads (34.3 percent), sliced plain (17.7 percent), and stewed or baked (5.7 percent). They took testimonials from women trained in home economics who declared the product's "goodness" and "ease of serving."[105] In what became a national trend in advertising, the sly flacks told the cooperative: "You need media that reach *women*."[106] This was mass marketing as sophisticated as that conducted by any industrial company anywhere in the country.

Permanent and prolonged consumption throughout the population—that is the aim the advertising. Don Francisco, the advertising mastermind at the California Fruit Growers' Exchange, told members of the industry that the point of merchandising was not simply to "educate" people about the benefits of eating oranges but to make them eat *more than they needed to eat*: "We can increase the demand for our agricultural specialties by making people want more. . . . Gradually we influence buying and consuming habits. We win consumers who never consumed before. We persuade former consumers to consume more."[107] The triumph of promotion was the creation of "a new generation," one that had been reared "on orange juice from infancy." In 1928, Francisco announced the results of twenty years of flackery for oranges and its sweet results. Consumption per capita climbed 72 percent from thirty-two oranges a year to fifty-five. More astonishing, the orange had become bigger than the apple: "The public may remember the slogan, 'An apple a day keeps the doctor away,' but it possesses much more specific and convincing information concerning the health value of oranges."[108] The J. Walter Thompson Company reported that per capita consumption of pears rose 70 percent from 1899 to 1925. Pears and oranges had ceased to be luxuries "purchased to gratify a whim or taste," and had become "almost staple articles of diet."[109]

Yet Don Francisco used this celebration of public relations to issue a warning. Agriculture could not regulate its output with the precision of manufacturing. Whenever the number of pears at market increased and prices again fell, the industry prepared for a new round of promotion. "Success always stimulates expansion," said Francisco. And in order to keep production from catching up with demand, "we must do everything possible to keep demand ahead of supply by proper merchandising and persistent educational work and then inform the industry of the facts on production when the facts predict overexpansion."[110] There is the suggestion here of an endless cycle in which advertising leads to demand that causes an increase in production that leads to new and more urgent campaigns. When they encouraged consumption for its own sake as a way to assure specialized crops a depression-proof market, growers committed themselves to eternal advertising and the constant danger of "underconsumption."

The Market Makes a Landscape

Orderly marketing could not proceed without imposing order on the product. Canneries and advertising brought pears to new consumers, but all consumption depended on quality. The public demanded fruit that approached in perfection the Platonic forms depicted in grocery store posters and popular magazines. To win their confidence and to provide them with high quality, growers of all varieties took up standardization.

The very idea of commodity pooling implied the classification of fruit into grades (for uniform quality) and standards (for uniform grades) in order that crops from thousands of producers could be transferred to the control of one marketing agency.[111] According to Herman Steen, editor of the *Prairie Farmer*, "Merchandising is virtually impossible without complete pooling, and pooling is thoroughly unsound without accurate and equitable grading. Pooling and grading are inseparable." Grades and standards represented nothing less than an attempt to impose industrial uniformity on fruit. "The pear grower is himself a manufacturer," said Frank Swett, who insisted that pears bearing the name of the association adhere to an ironclad set of criteria.[112]

Before grades and standards, fruit piled into eastern markets "regardless of quality, regardless of quantity, regardless of demand and regardless of all considerations that should influence intelligent business transactions."[113] Growers picked for size regardless of maturity. The orange growers of Tulare county shipped 20 to 25 percent of their crop "before the oranges were fit to eat." The produce reached a public eager to sample it, but subsequent shipments sold at a loss.[114] Without grades, buyers assigned designations of quality in the auction room where they opened boxes at random to determine firmness, shape, color, and size. Sound fruit sold high and blemished fruit sold low. The buyer determined the "grade" by agreeing to a price.[115] If the quality of California fruit remained unpredictable from box to box, buyers stopped recommending it to retailers, and shoppers on the street looked fondly on oranges from Florida. Said one speaker to the members of the fruit growers' convention in 1915, "The consuming public have been forced to realize that fruit shipped from California is not always the ripe juicy article that it is advertised to be." The damage that second-rate fruit inflicted on the trade could be serious and long-lasting. One distressed grower argued that a single inferior box affected buyers in the same way that "a single counterfeit note will cast suspicion upon a whole package of genuine bills."[116]

The cooperatives strained to convince wholesale buyers and shoppers that each piece of fruit was as good any other. When the harvest arrived at the packinghouse, workers threw it into bins where conveyor belts carried it to a sorting area. Trusted employees, and often the owners themselves, selected out the different grades according to size, shape, color, surface damage, and texture. Growers made these evaluations while packing in order to free buyers from suspicion at the time of sale. Said one reformer, "We must be guided by the experience and adopt the practices of other successful manufacturers, and so arrange and classify our products that each purchaser may secure the *identical commodity* he orders in the most convenient form to meet his requirements with out [*sic*] danger of adulteration or deception."[117] Grading created confidence in the product and encouraged faster and more specialized sales. Once accepted, they divvied up the trade into categories like *fancy, choice, standard*, and *pie* or *juice* grades.[118] The top of the line in the pear business was the No. 1 Bartlett canning pear. It measured 2¼ inches round after the crown and before the base, showed no flaws, displayed an even yellowish color, and looked like a pear should look. Pears that made the No. 2 grade held a perfect shape, but measured smaller than 2¼ inches. Fruit with "black end," a kind of blight, never made it into a box or can.[119] The baker could now buy pears and pay a lower price than the general grocer from a working-class neighborhood, who in turn paid less for them than the snooty buyer for the Hotel Astoria who wanted only the best.[120]

Still, the apparent order that grades represented needed some greater authority to stand behind them: the state. Only some centralized authority could decide for all producers whether or not a No. 2 pear was in fact a No. 2 pear. State governments began to impose standard grading laws as early as 1910, when Maine regulated the quality of its apples.[121] By 1917, most of the fruit-producing states had laws like California's Fresh Fruit, Nut, and Vegetable Standardization Act of 1917.[122] Its most conspicuous feature gave the state commissioner of horticulture police power to enforce all its provisions. This "commander-in-chief" could inspect any packing shed and warehouse in the state without notice, and he or she retained the right to seize and hold any fruit for evidence.

Here is the chief in action. When bad pears found their way across the country Frank Swett heard about it. The matter was serious enough for the horticultural inspector of one neighborhood to call on the pear association's executive committee to inform them that the laws of the state did not allow for the sale of wormy fruit to the general trade in San Francisco or anywhere else, and that unless they cleaned up their act right away the fruit would be quarantined or returned at the growers' expense.[123] After 1917 the federal government established its own inspection service through the USDA to resolve disputes between receivers and shippers over quality.[124] Few complained. For people whose fortunes

bloomed or weltered with the cycles of the greater economy, quality became the ultimate value. Not only did standards and grades uniform the
fruit, they protected the value of real estate, the security of loans, and the
cooperative itself as an institution with the public's trust. Standards and
grades became another safeguard of the specialized countryside.[125]

There is one last turn for us to consider in the career of uniformity,
and it provides our link between the search for order that began with
Harris Weinstock's journey and the landscape of the irrigated colonies.
Growers enlisted nature in their quest for quality by applying the standards of the packinghouse to the orchard itself. Indeed, it was a simple
step from standardizing grades to planting those varieties that manifested
the qualities of the grades. Their own depiction of fruit as immaculate
led growers to select a set of species with specific genetic characteristics
in order to sell the highest possible percentage of their crops as No. 1
Fancy. Good-looking fruit sold better than good-tasting fruit, so orchardists learned to plant the blushing grape, the roseate apple, and the
shapely pear. At least one advocate of standardized grades said that they
encouraged "pedigreed stock for future orchards" and helped to determine the "special fruits that should be planted in each district by the
elimination of . . . varieties of fruits that will not come up to the standard."[126] Reaching the market intensified the specializing tendencies of
commercial horticulture even beyond attention to natural advantages.

Growers began to farm with the knowledge that their products would
be judged on the basis of appearance—that the very images they helped
to create in the mind of the public would be used to evaluate the less
glamorous articles that landed in the produce section. The proclivity of
urban consumers to choose what looked good without regard to other
aspects of quality (like taste and nutrition) was apparent as early as 1904
when a writer for the Bureau of Statistics made a series of trenchant observations. George K. Holmes saw the future, and it was bland: "The
growing, the preparing, and the marketing of many of the products of
the farm are becoming questions of art and psychology. Less do people
eat to live than they live to eat, and yet when they buy food, they buy it
often not primarily for the gratification of taste, but upon the testimony
of the eye, which is pleased with form and color."[127] Holmes maintained
that "[i]n the city, the appearance of an apple is everything and taste
nothing."[128] He reported two recent competitions for the excellence of
apples both of which based over 80 percent of their criteria on appearance and almost ignored taste.

Holmes gave farmers a lesson in how to survive in the flash and splash of the city market. Consumers have a fondness for things red, he advised, and for yellow "when reinforced with large size." "Gloss, polish, and luster are wanted. Things should be large, and, when applicable, of plump appearance; they should be uniform in size, shapely, and with ornamental lines. A convenient name, once established favorably, catches the fancy of customers often more easily and securely than anything else."[129] Concluded the author, "Taste is the fruit grower's principal test of an apple, if he has to eat it himself, but very different attributes are of chief importance when he considers consumers in general, most of whom are townspeople."[130] Advertising explains only part of the banality of much modern food. People who had lost touch with the land idealized its products and pursued their ideal. Growers increasingly obeyed food fashion.

Pear growers considered the desires of townspeople and planted the Bartlett. Accounting for between 80 and 90 percent of the pear crop, the Bartlett became the most common variety in California.[131] Its popularity derived from "its early, regular and heavy-bearing tendencies, very good quality, unequaled drying and canning characteristics, good shipping qualities and wonderful adaptability to the varying soil and climatic conditions of the state." In other words, the Bartlett seemed made for the trade. Growers could get it to market before the pears of other regions, it could be sent in a number of different ways depending on conditions, and it had a good chance of surviving the journey.[132]

The great assortment of pears in cultivation across North America makes the overwhelming use of the Bartlett in California more striking. Americans knew sixty varieties of pears including the Bosc, Comice, February Butter, Giffard, Idaho, Joan of Arc, Le Conte, Onondaga, Sheldon, Urbaniste, Vicar, Wilder Early, and Zoe. Of course, the essential question for any California fruit grower was which of these sixty varieties could be grown to maximum advantage in some part of the state. George P. Weldon, author of "Pear Growing in California," a report for the state Commission of Horticulture, tested twenty-five varieties in eight subregions. Of these, fourteen tested "best" or "very good" (the only designations that mattered to Weldon's readers) for one or more of the regions. The author made special mention of the Winter Nelis as a commercial producer: "In California there is no better pear on the market, during the holiday season, than the Winter Nelis." Grown locally for sale in San Francisco, these pears fetched $2 to $3 per box in December of 1916. But the Winter Nelis fell out of favor.[133] The Bartlett grew at the top level in seven out of eight regions—a sure bet for anyone interested in putting in pears.[134]

The same process of selection narrowed the offerings of fruits all over North America. Two oranges made up 90 percent of California's crop. New York produced one hundred varieties of apples, but only ten were considered commercially important. Wrote the U.S. Department of Agriculture, in the old days "the producer set out a miscellaneous planting in order to offer his customers a number of varieties and to extend his marketing over the longest possible season." But with standards and grades, "many unsuitable varieties have been eliminated."[135] In this way, the landscape became calibrated to the harsh expectations of railroads and buyers.

That calibration even extended to the timing of key events in the seasonal cycle of the trees on the ground. The Bartlett protected growers from competition with producers in other regions because it came to maturity considerably earlier than pears in eastern districts. Bloom took place in the Sacramento Valley on or about March 25 and on April 9 in the foothills. Picking began on the river toward the end of June or the first of July and lasted to the middle of August. Under this schedule the first of the crop could go fresh to market a full month before pears from the Northwest, New York, and Michigan. By late August, California producers abandoned the fresh market to the East and put the rest of their crop into cans. When Pacific pears "were well out of the way" before the eastern regions shipped, everyone prospered, but the synchronization did not always work so well. In 1921 eastern supplies came out three weeks early, resulting in lower prices for everyone.[136]

When it functioned, this managed cycle coordinated the agricultural regions of the nation. Oliver Baker said that the maximum utilization of every climate and soil type would safeguard farmers against ruinous competition. But where the single crop appeared to solve certain problems it caused others. The narrowing of species seemed to promise control over nature—a way to strip down the complexity of ecosystems to just a few elements that could be understood and more easily manipulated. The perfect pear that consumers beheld in the pages of popular magazines depended on a separate logic of external controls. Mastery over the greater ecosystem, especially the containment of insects, became necessary as a consequence of the same genetic simplification that helped growers gain control over the market. What made sense in the calculus of commerce made for a risky ecology.

Figures 1 and 2. Growers. Charles Collins Teague and Harriet Williams Russell Strong exemplify the class of shrewd entrepreneurs who built the fruit business. These and other well-capitalized growers founded shipping companies and cooperatives; they called on the University of California to conduct research into pest control; and they established organizations like the Farmers' Protective League to promote their political interests. The seasonal pattern of fruit cultivation—a hurried harvest followed by months and months when the trees demanded little attention—gave growers the time to build an industry. Photograph of Teague reprinted by permission from Alan Teague, Santa Paula, California; photograph of Strong courtesy The Huntington Library, San Marino, California.

SOROSIS FRUIT COMPANY.

SOROSIS FRUIT COMPANY.

Figures 3 and 4. Packinghouse and auction house: two sides of the fruit economy. Packinghouses like this one focused the crops of hundreds of orchards and enabled growers to make market connections. But the integration of California into the national economy could not proceed as long as marketing remained in the hands of commission merchants and others with little experience in buying and selling fruit. The auction moved produce faster from the arrival yard to wholesalers, but the system still kept growers outside the process of distribution. Both photographs courtesy The Bancroft Library.

The California Pear Grower

Vol. IV.   SEPTEMBER, 1924   No. 8.

IN A NEW YORK FRUIT AUCTION.

"I take mine this way"

Food for thirst on a hot day—fresh California Bartlett Pears! How good it feels to your parched throat—the cooling, refreshing juice of this luscious pear! That's Nature's way of quenching thirst. A fine way, too—cools you off and doesn't spoil your stomach.

So where's the *summer-sense* in loading up on bulky foods and icy drinks when these meaty, juice-full pears give you food-and-drink in one?

At the nearest fruit stand pick up a handful of these fine, fresh Bartlett Pears and let Nature "mix" your summer-day drink. While you're at it, better buy a dozen. There are always plenty of thirsty folk around.

CALIFORNIA PEAR GROWERS ASSOCIATION
SAN FRANCISCO—CALIFORNIA

fresh CALIFORNIA

BARTLETT PEARS
— *buy them by the dozen*

Figure 5. Promoting the perfect pear. Unlike most other agricultural commodities, fruit could be offered to the public as a finished product. When the canned fruit market failed in the early 1920s, growers turned to the new art of "consumer education" to sell their surplus. Advertising encouraged consumption for its own sake and gave fruit an image, an ideal, and an attitude. But although promotion may have saved thousands of acres of orchard trees from being torn out in the 1920s, it also set a high level of sales that growers, henceforth, had to maintain. Courtesy The Bancroft Library.

Figure 6. Spray chemicals. Monoculture caused an explosion in the population of insects. In one sense, fruit growing became an industry when business enterprises like spray-chemical companies furnished growers with manufactured inputs. Note the labor necessary to apply these products, as well as the possibility of exposure. Though chemicals usually paid for themselves in saved fruit, they also contributed to the rising cost of fruit growing. Courtesy The Bancroft Library.

Figures 7 and 8. Two portraits of labor: Asian fruit pickers and white
European fruit pickers. The labor needs of orchards and vineyards came
all at once, usually within a period of a few weeks each year. Few growers kept hired hands all season long, so a market in temporary labor
emerged in which pickers moved from place to place, following the crops.
But growers claimed in the 1880s that Chinese immigrants should not
be given orchard work, that only white workers should do the picking.
In time, employers favored any group poor and desperate enough to
take low wages and hard conditions. Courtesy The Bancroft Library.

Figure 9. Orchard home, 1890s. "A little land well tilled" describes the property depicted here. Perhaps Doctor Strentzel retreats to his Alhambra Ranch
on weekends while a manager takes care of things during the week, or perhaps
he has retired to this shaded valley. Yet the orchard idyll depended on intensive cultivation—a costly form of agriculture. Rather than farm fresh land with
old methods as a way of raising yields, growers added capital, technology, and
labor in order to harvest as much product as possible. Only big harvests paid
for the high costs of production and sale. Courtesy The Bancroft Library.

LANDFORMS OF CALIFORNIA
and NEVADA

Second revised edition.
1951

from W.W. Atwood: *Physiographic Provinces of North America.* Ginn & Co. Boston Mass.
by Erwin Raisz

Scale 0   100   200 Miles

MAJOR LANDFORM
REGIONS

Map 1. Valleys and rain shadows. California's boundaries just happen to define a Mediterranean climate with many subregions, like arid interior valleys and damp coastal plains. A diversity of climates and soils encouraged a diversity of specialized crops. After years of experimentation, a pattern of districts emerged that defined where certain fruits could be harvested in quantities that paid for the many costs of intensive agriculture. Copyright by Erwin Raisz, 1965. Reprinted with permission by GEOPLUS, Danvers, Massachusetts.

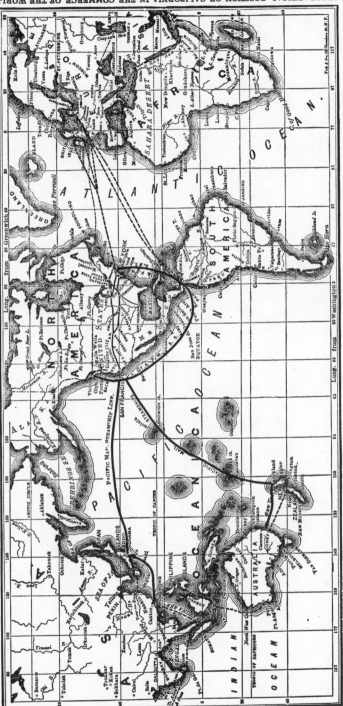

Map 2. Relative advantages, 1880s. Boosters imagined California at the very center of the world's wheat trade, despite the state's obvious isolation from market cities in North America and Europe. For twenty years merchant-farmers maintained a far-flung commerce in grain. Yet even the proposed route through the Isthmus of Panama and a steady communication with Canton would not have forestalled the decline that followed when farmers in other, better situated regions sent their wheat to the same markets, bringing competition and lower prices. By the early 1880s the risk no longer paid and the wheat trade died. Fruit growers too learned that the benefits of natural advantages could be destroyed by risk, insects, competition, and anything else that caused high costs. Reproduced from the *Pacific Rural Press*, Courtesy The Bancroft Library.

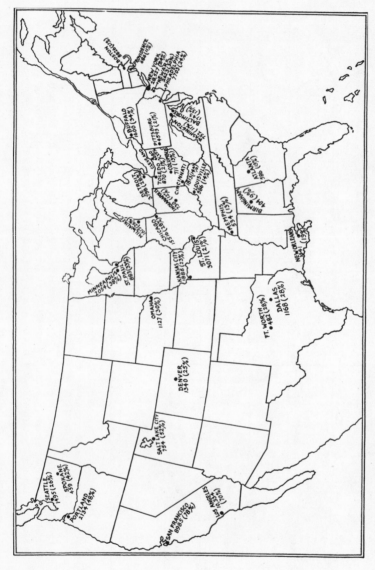

Map 3. California's market cities in 1924. The numbers represent unloads of California fruits and vegetables in each city, along with the state's contribution to the car-lot supply. New York's truck gardens and orchards had furnished city people in the East with peaches for almost a century by the time irrigation became widespread in California. But by the 1920s New York had long since lost the fruit advantage. Courtesy The Bancroft Library.

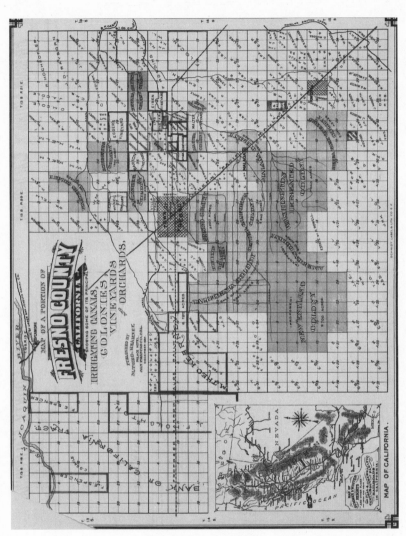

Map 4. Colony lands: Fresno County. The Central Valley became a region of small irrigated farms. Land developers, operating in companies like the Washington Irrigated Colony and the Bank of California Tract, subdivided great acres into modest tracts and sold them with a water right and a promise of country repose. Note the irrigation canals and the tracks of the Southern Pacific Railroad—critical elements in the transition of this landscape from extensive to intensive cultivation. Courtesy The Bancroft Library.

4

A Chemical Shield

The pursuit of fruited wealth in the markets of the world brought ecological disorder to the land. Trees and vines covering thousands of square miles of once diversified grasslands, foothills, and riverine swamps amounted to a radical simplification of these natural landscapes. A few cultivated plants replaced hundreds of animal, plant, and insect species over hundreds of square miles—a transformation not without consequences. This simplified ecology invited a population explosion among alien (nonnative) insects. At the same time growers planted colony tracts and attempted to reach markets abroad, their orchards fell under attack by insects so perplexing that they sometimes evaded identification for years, insects so resilient that no existing practice or poison stopped them for long. The precision that cooperatives brought to the selection and sale of fruit belied an ecology seeking stability in chaotic circumstances. The greatest threat to the specialized districts was the environmental consequences of specialization itself.

In the midst of an ecological and financial calamity, a network of scientists, agencies, and industries came forth to rescue fruit growing from the effects of its own expansion. Growers with limited knowledge and funds could not combat scores of pests on their own. Between the 1880s and World War I an agricultural service industry centered on insect control emerged to furnish information, regulation, and chemicals to the orchardists. From the University of California came the Agricultural Experiment Station, a fountainhead of research and leadership. The state of California established a regulatory police force to quarantine infested orchards, and state commissioners and entomologists, some of whom opposed the use of chemical insecticides, offered beneficial or predator insects as an alternative form of pest control. Under the auspices of the first two institutions came a third. By grafting the chemical technology of the University with the regulatory force of the state, the spray-chemical industry became the principal dispenser of insect protection in the twentieth century.

These institutions protected specialized agriculture against nature's response to a simplified environment. Insects had the capacity to ravage the fruit business out of existence by lowering yields and destroying quality. So any tool that enabled growers to combat the plague became essential to the very process of cultivation. Just as the automobile called into existence glass, steel, and rubber companies, so too did commercial horticulture depend upon other industries to furnish materials for the assembly of the final product.[1] The lovely pear of the shopper's ideal was as much the product of chemical research as tight standards. The output of the agriculture service industry became as material to the growing of peaches and pears as irrigation water, soil, and sunlight. Spray nozzles in the hands of thousands of small-scale growers became implements in the creation of a new rural landscape—an ever-expanding region of specialized crops protected by a million-dollar chemical shield.

Single-Crop Ecology

To mark the ecological changes that commercial horticulture brought, it is well worth a short jaunt to the tules. Tule, a long reed-like grass, a kind of bulrush, once grew in the lowlands and marshes that covered vast expanses of the Central Valley. Unlike the domesticated trees and vines of the cultivated landscape, tules did not live in pure stands apart from other plants and animals.[2] They held a place in a much more diversified ecosystem, one composed of valley oaks, elk, and insects such as butterflies, dragonflies, mosquitoes, and webbing spiders. When John Muir crossed the Pacheco Pass and entered the San Joaquin Valley for the first time in April of 1868 he discovered a landscape almost unrecognizable from the "desolate waste" that William Brewer suffered three years before. A season of rain ended in a shocking blue sky and long views to the mountains, unified in the middle ground by miles of windblown flowers. Within one square yard Muir found ten orders, sixteen species, along with abundant native insects, and enough inspiration for a reverie: "In this botanist's better land, I drifted separate many days, the largest days of my life, resting at times from the blessed plants, in showers of bugs and sun-born butterflies, or I watched the smooth-bounding antelopes, or startled hares, skimming light and swift as eagles' shadows."[3] Grasses and sedges like tules compared to wildflowers in their variety. Before Spanish missionaries and American ranchers brought cattle to graze the bottom-lands, perennial species dominated the Central Valley. Purple stipa, nodding stipa, California melic, California brome, and Junegrass were only a few of the long-lived bunchgrasses that covered the valley floor. When Europeans brought more aggressive annual grasses (most by accident, some by design), the new arrivals displaced many native species. One especially successful introduction, wheat, eventually shut out all competitors, and the rush for orchard tracts sent surviving native grasses to the corners of the land—to fence posts and foothills. Over one thousand species of grasses covered the hills and lowlands of California before the events of the nineteenth century.[4]

Diverse plants imply diverse insects. The two evolve together, with certain insects breeding and feeding in association with certain plants.[5] Some feed on leaves, fruit, and roots, while the predacious kind feed on other insects. What rules this relationship and keeps it steady over time is the food chain. Plants consume sunlight, water, and carbon dioxide; insects consume plants; birds and mammals consume insects (as well as plants); and the larger mammals feed on the smaller. Ecosystems appear to work in a self-sufficient, self-regulating manner because over long periods they reach a moving equilibrium among their elements. Wild plants rarely disappear in a cloud of bugs, because their leaves and bark are toughened against herbivore advances.[6] And the presence of many different plants means many species of insects, providing enough predators and prey to keep the total population in check. There can be only as many predators as there are prey to sustain them and only as much prey as the edible plants and the predator population allow. This tension remains until or unless some element is thrown out of the system. In this event, insect populations seek a new equilibrium.

Farming radically simplifies the genetic diversity of any natural landscape. The cultivation of certain plants to the exclusion of all others is just what it means to raise food. Weeds (undesired plants that impinge on and sometimes attack cultivated plants) and pests (insects, mites, birds, and mammals that do the same) are as much a creation of human culture as they are of nature. There are no weeds or pests in a tule swamp; we reserve these classifications for organisms that get in our way. Animals and plants become pests when they prosper in humanmade environments; and in fact, certain species are so well suited to backyards, waste places, and fields that they would scarcely exist without these human landscapes.[7] Insects, in particular, flourish on the plants that people have propagated for nutrition often at the expense of immunity to herbivores. Food plants are thus in constant need of human attention and protection. If over the centuries people did not develop a vigilance in defense of their crops, neither would have prospered. Pest control in some form is as old as agriculture.[8]

Excessive insect populations resulted from the selective cultivation of a few crop plants over large regions. Any region in which the flora is simplified by agriculture or for any other reason presents an open invitation to plant-eating insects to move in and exploit the niche. Lacking predators and competitors for food, a single species will proliferate, limited only by the food supply.[9] The consequences of simplification were even more severe in California where the trees and vines on colony tracts came from around the world. Nurseries imported orange trees from Australia and Brazil, grapevines from France, and apple trees from the eastern seaboard. Between 1810 and 1942 close to 570 species and varieties of trees—not shrubs or bulbs or vines or flowers, but trees alone—arrived in California. Most of these entered between 1860 and 1920, along with 260 species and varieties of vines—including flowering and decorative vines as well as grapes.[10] Exotic varieties included rare apple trees, cherry, plum, and prune trees; quince, olive, and Japanese persimmon; guava, loquat, medlar, and mulberry; gooseberry, blackberry, and dewberry in many varieties, such as Crandall's Early Blackberry or Smith's Improved Gooseberry or Loangworth's Prolific Strawberry. Not all of these came from continents other than North America, but none were native to California.

These plants brought legions of insects nestled in fruit, hidden under leaves, or snug beneath bark. Back in their native environments these bugs had to contend with competitors and predators that governed their numbers. But upon arriving in western North America, fruit pests entered a land with few enemies, a garden rich in food, a New World. Nothing short of dramatic human intervention could stop the exponential increase of exotic bugs as they fed on the leaves and fruit of their favorite plants without regulation. The landscape represented by tule reeds came to be replaced by one composed of a handful of plants and the insects that plagued them. Before fruit crops became important to the state's economy farmers could raise food with few pests, but by 1885 a full-scale invasion had hit the state, and growers cast about for solutions.[11]

The enemy was ubiquitous and formidable. Pear and quince trees were known to host fifteen different pests that attacked their leaves, buds, flowers, and fruit. Grapevines had fourteen known enemies including San Jose scale, grape scale, black rot, flea beetle, berry moth, and leafhopper, as well as phylloxera, which attacked the roots.[12] Blister mites attacked grapes, walnuts, pears, and other trees by eating the leaves, causing them to become distorted, turn brown, and fall off. Pear and orange thrips ate the bud and the embryo fruit, and they often reproduced deep in the interior of the host plant, making poisoning nearly impossible. The many other insects laying eggs in groves up and down the state had names that growers learned to fear: woolly apple aphid, codling moth, cottony cushion scale, red scale, red spider mite, and Hessian fly—all introduced into California since the Gold Rush and all economically threatening.[13] Said one exasperated grower, "We have the pests of the entire world, and from present appearances they have come to stay."[14]

The insects could not be controlled. Infestation of difficult-to-locate pests like the peach tree borer could destroy up to 50 percent of an orchard in a single season.[15] A vineyard might lose money because of a shortened yield or because the crop arrived at market bug-ridden and rotten. Even a slight amount of insect damage hastened decomposition during transit, and wholesale buyers did not find half-eaten pears attractive. The secretary of the state Commission of Horticulture estimated that insect damage amounted to a 10 percent tax on all fruit products.[16] Matthew Cooke, a fruit box manufacturer from Sacramento and an early proponent of state activism in insect control, made his own survey in 1881 and concluded that only 35 percent of the harvest was fit to be sold. The damage amounted to $2 million—a sum equal in value to the entire crop of 1880.[17] No industry sustaining such losses could exist for long.

Amid the panic to cleanse their trees and vines, growers ruled out certain options. Crop diversification as a means of encouraging predator insects received little attention. No orange grower harvesting up to one thousand boxes per acre would have considered turning over a portion of his or her land to walnuts for the sake of ecological diversity. In any case, a grove of walnuts and orange trees does not amount to polyculture, which implies the distribution of resources and labor over a number of crops as well as a degree of self-sufficiency from the market. For most of the insects doing damage there were no predators to encourage anyway. Turning the trees under by burning and replanting was too costly to contemplate. A wheat farmer might destroy an infested crop and only suffer the loss of a season's income, but orchards represented thousands of dollars and years of work to bring to maturity. And what would keep the insects from plaguing the new trees? Faced with such bleak prospects, growers looked for alternatives that would rid them of insects while leaving specialized horticulture intact.

What follows is a story of alternatives. Chemical insecticides, some already centuries old in the 1880s, some representing new technology, emerged as the weapons of choice among fruit growers. The rise of chemicals did not go uncontested by those who recognized their flaws, but chemicals succeeded. Bug spray carried consequences less obvious than poisoned soil or poisoned people; rather, it summoned new sources of private and public authority.

Mixing Chemicals

Pest control grew up with the fruit business. As long as production remained isolated in islandlike districts, outbreaks of destructive insects did not spread across the state, at least not very quickly. Some plant maladies went unidentified as insect damage for years before anyone began looking for a cure. One such pest, phylloxera, evaded identification until 1873, after clandestinely killing off grapevines for fifteen years.[18] The University of California, founded in 1869, lacked the funds, facilities, and faculty to rescue panicked growers. Instead, orchardists fended for themselves with a crude collection of powders, oil soaps, and petroleum washes, as well as sulfur and tobacco, which they sprayed, sponged, and sprinkled to fortify their crops. To kill pear and orange thrips, mix one part black tobacco leaf with sixty parts water; combine this with a 2 percent solution of distillate emulsion (an oil and water mixture). Then prepare another formula consisting of twelve gallons of hot water, thirty pounds of whale-oil or fish-oil soap, and twenty gallons of the previously prepared tobacco and oil solution. Dissolve the soap in the hot water and pour it into a spray tank. Add the oil and spray at high pressure. If the mixture failed to kill the thrips there were other recipes in other manuals. Growers learned to test any plausible mixture in a race to exterminate the pests.[19]

None of these cures actually eliminated insect populations. At best they acted as temporary deterrents. Considering the ignorance of the people who used them, it is amazing that these remedies did any harm to the insects at all. Sometimes growers sprayed or washed the orchard after the bugs had done their damage and laid their eggs; the eggs then survived to hatch later in the season or the next year. Knowing nothing about insect physiology, orchardists routinely chose the wrong application for the trouble at hand and often misused the chemicals.[20] Broadly speaking, there are two kinds of herbivorous insects classified in terms of how they feed: those with horizontal mandibles that eat through leaves and fruit and those with beaks that poke through the skin. The first kind can be killed by poisoning the surface of their food supply. But the long-beaked species must be killed by contact with a corrosive substance that makes them unable to move or breathe.[21] Growers commonly used the wrong chemical or succeeded without knowing why. One orchardist brewed a solution of lime, sulfur, and salt to kill San Jose scale but had no idea which ingredient actually did away with the bug: "Some experiments have been made to find out which of the materials is the insecticide, but it cannot be claimed for either separately, but to the combination."[22] Other recipes concealed similar mysteries, and successful growers became avid experimenters into the efficacy of various common chemicals, bringing many mixtures to the attention of the state university in Berkeley.

When growers looked for stronger stuff and expert advice they increasingly looked to the University of California. Meeting in convention in 1883, farmers from across the state demanded that the university direct its attention to the insect disaster. They called for a campaign of public education and a chair in entomology at the university: "We want the very best man that can be had, . . . [one] who is in the prime of life, so that he has many good years of valuable service to give us, and who will come here with the understanding that he is to be an instructor in entomology as his great work."[23] For the time being, however, they had no one but a soil scientist to help them. Eugene Hilgard and his philosophy of practical solutions landed in California just in time to conduct the state's first full-scale pest eradication program: phylloxera of the vine.

Phylloxera vastatrix, an insect the size of a grain of sand, lives on the roots of grapevines, where it eats away at the water-absorbing tissue. The leaves turn brown, and the plant dies of dehydration. A map dated 1880 shows the phylloxera threatening the state's leading viticultural regions: the wineries of Sonoma and Napa Valleys and the table grape vineyards near Placerville in the Sierra Nevada.[24] The tiny bug presented such a problem because it lived underground and thus out of reach of whale-oil soap and insect powder. Hopes that an introduced predator insect, the phylloxera thrips, might control the scourge came to nothing. In 1884 the College of Agriculture reported that "the list of bootless propositions" had steadily increased as self-proclaimed experts came forward with experiments. Flooding the vineyard to drown the phylloxera helped in some cases; heavy fertilization worked in others, but no pattern came clear.[25] Exposing the roots for fumigation by removing the soil meant giant labor costs.

Chemical cures appeared to have great potential. Simply mix some sulpho-carbonates into the irrigation water and let it sink in. But these treatments required more than one application, they were expensive, and they often failed to reach all the infested areas. One terrible idea that briefly commanded everyone's attention was mercury vapor. The method was suggested in 1885 by J. A. Bauer, a grower who diffused finely divided quicksilver into the soil around the stems of his vines and claimed to have killed the phylloxera. The state Viticultural Commission performed its own tests at sites in the Napa Valley, and the university hurried to learn Bauer's methods for its own investigation. The vines survived unharmed, but so did the mighty phylloxera.[26] Although other tests showed better results, none proved decisive.[27]

Finally in 1886 Hilgard performed experiments with roots from the eastern United States, where phylloxera lived among wild vines without causing injury. The phylloxera could not be exterminated, but its effects could be neutralized by grafting cultivated European vines to resistant roots.[28] The phylloxera continued to be a problem in grape-producing regions, but Hilgard's solution put to rest many fears and earned the university praise. The staff at the experiment station emerged with a sense of their ability to undertake research and to inform a broad and literate audience willing to put findings into practice. The sense of triumph, however, did not last for long. Grafting resistant rootstocks saved the vines, but the procedure still required growers to tear out their vineyards—just what they had hoped to avoid. More important, though the university had proved itself equal to the task of solving a genuine insect mystery, Hilgard's experiment station had spent four or five years trying to stop one insect. In the meantime, California had become the promised land for an entomologist's catalogue of other bugs, all of them equally damaging.

Phylloxera was by no means the only affliction in the land. Long before it ceased to be a primary research goal at the university farm, its reputation had already been eclipsed by many other insects, all wreaking havoc at the same time. In 1882 a commissioner with the recently created Board of Horticultural Commissioners warned of impending ruin: "I have visited all the orchards in this district. Three-fourths of the entire crop of this season has been destroyed by codlin moths."[29] Another commissioner told of red scale in Southern California's citrus orchards, black scale in the Central Valley, San Jose scale in Santa Clara County, and red spider throughout the state. No one insect could command the university's attention, and no one institution could take on the hordes alone. "It will be readily seen," concluded the commissioner, "why a united warfare on insect pests . . . should be organized."[30]

State Business: Orchard Police and Biological Control

The phylloxera did more than awaken the university, it also awakened the state of California to the fact that agriculture had emerged as its next major industry.[31] Mining was clearly on the decline in the 1880s, and the Southern Pacific Railroad wanted to sell land along its route through the San Joaquin Valley and Southern California to actual settlers who would support the road by shipping their produce to San Francisco and Los Angeles. Growers appealed to the state legislature in Sacramento to do for them what they could not do for themselves: control bugs through quarantine legislation and enforcement. Infested cuttings and fruit boxes moved from region to region, transplanting insects and eggs along the way. Since individual growers and the railroad refused to modify their practices for the common good, regulatory force became popular. The Horticultural Quarantine Law of 1899 established officers and guardians to inspect all shipments and seal the borders against the insects of the world.[32] In time, the state sponsored, in addition to regulation, a research program of its own into predatory insects and purification laws for insecticides. The expanded role of the state followed, not only from a desire on the part of the legislature and the governor to enhance the gross product, but from the urging of growers, one of whom declared: "Our legislature should step up to a higher plane of duty and pass such laws as will render possible the protection of our orchards."[33]

The state became active in the fruit business in 1880 when it established the Board of Viticultural Commissioners. The new agency was to report on the advance of the phylloxera and diseases of the vine and work with the university to educate growers. Prominent growers like R. B. Blowers, the Sacramento Valley shipper, and Charles Krug, a leading vintner from Napa, sat on the board along with Charles H. Shinn, who served as commissioner at large.[34] Still, this committee hardly formed the long arm of the law that many hoped for. In March of 1881 the legislature responded to complaints of inefficacy with an act creating a state Board of Horticulture (later the state Commission of Horticulture) with the power to enforce quarantine regulations. Appointed by the governor, the six-member board had the right to enter any orchard they pleased at any time and inspect "stock, tree, shrub, plant, vine, cutting, graft, scion, bud, fruit-pit, fruit, vegetable, or other article of horticulture, or implement thereof" and recommend action.[35] Owners who refused the commissioners' authority had their trees treated against their will, even to the point of burning them to protect the property of others. In later versions of the law each county established its own county board to scout the territory and report news of infestation to Sacramento.[36]

Fruit growers could not help but be equivocal about the state's "protection" of their trees. In 1886 portions of Fresno County suffered an onslaught of San Jose scale, a bug that attaches to branches and trunks. When the inspector lost his faith in coal oil, he cut off the infested areas from all commerce. The state demanded that all local fruit be examined before leaving the areas and that no more imported trees be brought in.[37] When all this showed no results, the inspector showed no mercy and ordered that the trees be burned to cinders to prevent further spread. The orchardists probably did not protest the solution, recognizing that their trees were already beyond help. But destroying bearing groves would shut down the fruit business before it rid the state of damaging insects.

Clearly, not all growers were excited by the idea of state intervention into their practices, but those who supported the commission were just as loud. C. H. Dwinelle, president and commissioner at large of the California Board of Horticultural Commissioners, defined the solution of the insect emergency as the combined efforts of self-interested growers. "The greatest dangers to the industry are now from insect enemies," said Dwinell, and he voiced optimism that the scourge could be fought with "vigilance and industry." The only remaining issue was the resolve of the growers: "The important question now is, who are to be one's neighbors? Will they breed pests for the orchards of the thrifty, or will they have intelligence enough to study the habits of noxious insects, and pluck to fight them?"[38] Matthew Cooke called for an end to the individualism that kept growers from acting in the interest of the industry as a whole. The chaos resulting from an attitude of "go as you please'" could only be restored to order, Cooke believed, "by the aid of legislation."[39] Interstate quarantine boards and purity laws regulated fruit shipped outside the state and the quality of insecticides used within. All these controls came at the request and to the relief of the great majority of fruit growers.[40]

The state and the university, however, did not always agree on means. State entomologists, acting in what they believed to be the best interests of business, supported the introduction of beneficial insects to fight pests without chemicals. At a time when the future of pest control had yet to be determined, introduced insects offered a clear alternative to chemicals. At issue was more than the manner in which orchardists would protect their trees, but the location of authority in questions of agricultural research. Both the state and the university had been handed power by growers. It was only a matter of time before they would come to odds about who should plan the future of industrial agriculture in California.

The state's most farsighted contribution to the fight against pest insects was its work on beneficials, or "nature's way" of fighting bugs with other bugs. Since so many insects causing damage in California came from elsewhere, why not go elsewhere to look for their natural predators? Beneficials emerged as the state's own research project and a near obsession among those who opposed the use of sprays.

In the most famous example of biological control, an entomologist working for the state demonstrated the low cost and stunning efficiency of a predator bug. Some time in 1868, a nursery in San Mateo County imported a shipment of lemon trees from Australia. Unknown to the owner of the nursery or anyone else, the cuttings hosted the cottony cushion scale, an insect that sticks to the branches and leaves of trees in white bunches. The scale quickly became one of the most destructive citrus pests ever to enter the state. Soon after, a Los Angeles nursery imported some of the same stock and introduced the scale to Southern California. Orange growers tried everything to get rid of it: alkalis, oil soaps, and new arsenic-based chemicals. Nothing worked because the scale's waxy skin shielded it from liquid poisons.[41] By the middle of the 1880s, the groves of Los Angeles, San Bernardino, and Orange Counties sustained an infestation so severe that observers likened the living cloud to a softly falling snow. Other varieties of fruit also declined, but oranges got hit the hardest. Growers who had once made five hundred to a thousand dollars an acre faced extreme losses.[42] Digging out and burning them was the only sure way to rid the trees of scale, and after replanting the bug simply came back.[43]

Tellings differ regarding who proposed the magnificent solution and when, but certain facts seem clear. A delegation sent to the Melbourne

Exposition of 1888 included a Californian named Frank McCoppin, who knew that horticulturists were desperate for help. Orange growers suspected that since the scale came from Australia and since the bug did not seem to be a problem for orange growers there, a predator must exist on the island continent. McCoppin's friends asked him for help in securing a professional entomologist to do the work. So he wrote to C. V. Riley, chief entomologist at the United States Department of Agriculture, asking him to recommend an entomologist who might come along to Melbourne.[44] The USDA selected Albert Koebele. Koebele sailed with McCoppin to Australia and came back with a species of ladybird (*Vedalia cardinalis*, also known as a ladybug) that he claimed would eat the scale. Unbelievably, it did. In some places, growers claimed the scale disappeared overnight. The fruit growers and the USDA demanded other such miracles, and the California legislature was so impressed that it appropriated five thousand dollars for a second trip.[45] Koebele remained in Australia for about a year and returned with a number of other ladybug species that, he assured a public still nervous about introduced insects, were only beneficial.

Proponents of beneficial insects insisted that the victory of the vedalia need not be a onetime event, but they warned that the success of "nature's way" demanded commitment to the cause. The chemicals under investigation by the university and individual growers did not offer long-term solutions, they argued, and could only be temporary: "The California method of fighting insect pests is to use the most efficient artificial means while we have to, and to this end we apply all sorts of known washes, dips, and fumigation, but, while so doing, we realize that these measures are very cumbersome, costly and inefficient, and that nature has provided a better way, and it is of this way that we avail ourselves. . . . It is the policy of this State to use artificial remedies so long as there are no better ones, but to secure, introduce, and distribute the better means, and these consist of beneficial insects, as soon as possible."[46] State entomologists presented the accomplishments of the vedalia and other beneficial insects as a critique of chemical insecticides; that is, as a permanent program and not merely a stunt. As for the unbelieving "scientific men" at the university, Ellwood Cooper, president of the state Board of Horticulture, castigated their silence and questioned their motives: "Scientific men, who are in a position to render great aid to relieve the farmer or orchardist, have disputed the efficacy of the natural method, whether from ignorance or jealousy remains to be determined."[47] There could be no mistake about the incompatibility of the two methods; ladybugs could not be introduced into orchards sprayed with insecticides—they would die as surely as the damaging insects.[48]

Cooper's words carried the gravity of his position: founding member of the state Board of Horticulture, a commissioner for five years, and president for twenty years. As an olive grower, Cooper had experience to back his assertion that chemicals, though effective and on the rise, could never stand as a permanent solution. Not only did they fail to halt the spread of insects over time, but they also represented a significant cost to growers.[49] He claimed that to spray his olive trees with kerosene emulsion cost him three thousand dollars a year, "and [I] was not getting fruit enough to pay the expenses." As late as 1913, with chemical solutions firmly established and in widespread use, Cooper continued to rail against them: "Any other remedy is expensive, has to be repeated, is unsuccessful and idiotic."[50]

Nonetheless, the project to unleash predators against orchard pests folded because it required the kind of patience that one could only expect from an entomologist. Orchardists all over the state heard news of the great vedalia and believed that other species like it existed somewhere, but they also knew that their crops might vanish in a swarm while they waited for results.[51] The introduction of two other bugs, the *Orcus chalybeus* and the *Orcus australasia* did not bring relief after three generations. "Their increase will be slower," wrote the author for the state Board of Horticulture, and he worried that the "unprecedented success of the *Vedalia cardinalis* has caused fruit growers to expect immediate and similar results from all the new insects."[52] In addition to this, introduced predators often did not survive outside of their native environments, and a plentiful source of food might not make up for traumatic changes in temperature and humidity. Perhaps most important, predators came to eat a specific pest. It did owners little good to know that the vedalia had the cottony cushion scale under control when they were being ravaged by red spider. Chemicals killed everything and offered instant satisfaction to the bug-weary. In addition to all of these reasons, chemicals had the advantage of being new, scientific, progressive.

Thirty-seven insects—each thought to be the enemy of a known pest—had been discovered by 1906, but it seemed to Ellwood Cooper and his colleagues that no one cared. Not Theodore Roosevelt, not state governors, not the cotton growers suffering the onslaught of the boll weevil—Cooper tried to convince them all that predators could rid the nation of damaging insects, but to no effect. He claimed that the secretary of agriculture would rather have cotton planters raise corn than employ a predator against the boll weevil. No bulletin on the subject issued from Berkeley before the 1950s. Summing up the university's position, Edward James Wickson, the editor of the *Pacific Rural Press* and a professor at the university, told a Farmers' Institute gathering in 1903 that "the work of beneficial insects is not only cheap, it is efficient beyond the best insecticidal treatment. Unfortunately California growers cannot yet fight all the injurious insects with their natural enemies." Wickson, too, looked forward to the day when biological solutions would replace chemicals. He believed, however, that that day had not yet come.[53]

The university, largely by winning the allegiance of orchardists to chemical cures, rendered itself the central institution in the matrix of industrial farming. Unapparent before the question of biological control surfaced but clear in its passing was that the university would determine the course of research and the state would patrol the orchards. The possibilities of predator species remained in the background until the 1950s, when Harry Scott Smith brought a research program to the university's experiment station at Riverside that was both old and new. Until then, chemists and engineers and company presidents influenced the direction of research. As the university became more deeply invested in chemical insecticides and as the memory of the vedalia's triumph in the citrus groves faded, the advocates of "nature's way" felt left behind in a fine mist.[54]

Scientific Authority

Growers wanted powerful chemicals and experts who could prescribe them. In 1891 Charles W. Woodworth (1865–1941), one of the leading economic entomologists of his time, arrived in California to become the first professionally trained pest-control specialist at the University of California. He was born in Illinois where he attended the state university. Upon finishing two years of graduate work at Harvard in 1888, he became a botanist and entomologist with the Arkansas Agricultural Experiment Station.[55] Woodworth demonstrated competence in mathematics, optics, physics, and chemistry, but he disdained science apart from application. Like Eugene Hilgard, he made no distinction between pure and applied science—all worthy experiments tended to some material end. "The large experiment," he once said, "is the final test of the practical value of the facts."[56] He was especially interested in fumigation and chemical poisons and liked to test his prototypes in the field. If a method of pest eradication worked on a dozen trees but failed on a thousand it was worthless to Woodworth. Being an economic entomologist meant studying the losses due to insects to help growers weed them out of their trees and out of their balance sheets. This was no bug-loving scientist. Woodworth rarely studied an insect for any other reason than that of finding the most efficient way to kill it.

As Woodworth settled in his new position he learned of a new chemical recently invented in Massachusetts.[57] Arsenate of lead, one of a handful of new synthetic poisons, would become the most important insecticide in the world before DDT. Its predecessor was Paris green, a pigment made of copper and arsenic acid most often used to color paints.[58] Paris green first came to the attention of farmers in 1867 as a way to fight off the Colorado potato beetle. The exceptional killing power of this new compound, and the spray pumps and long applicators invented to dispense it, promised to change forever how farmers and growers managed insects. Then in 1892 Paris green did almost as much damage to the New England countryside as did the Gypsy moth, and some warned of its hazards.[59]

Poorly understood and potentially dangerous, arsenicals brought new perils as well as new hopes. To explain, the arsenic active in arsenical poisons was not the gray metallic element, the common poison of centuries past, but a compound called white arsenic (arsenic trioxide) also known as arsenic acid.[60] This acidic chemical reacted with a metal like copper or lead in such a way that it formed a tiny grain and a fine powder.[61] The grains sat on the leaves, buds, and fruit of the tree where insects ingested them while feeding. All arsenicals carried one giant caution: to be safe for the trees they could not contain too much arsenic acid. When an excessive amount became uncombined or "free" from the lead it burned the leaves on contact, defoliating the tree.[62] The water solubility of this caustic poison made the conditions of its use critical. Dense fog or high humidity might increase the availability of otherwise safely combined arsenic acid, causing burning that would not have occurred in dry weather.[63] When a chemist in Massachusetts came up with lead arsenate in 1892 he called it gentler on foliage than Paris green, but in fact it could do just as much damage in damp conditions.[64] The appearance of chemicals like arsenate of lead—a more complicated and powerful toxin than tobacco or whale-oil soap—signaled that killing insects would increasingly require the aid of experts like C. W. Woodworth, who understood chemistry and insect physiology and who could recommend to growers the best and safest chemical for their specific problems.

Chemical research in California did not begin with Woodworth, but he brought it purpose and direction. At the time of his arrival the university seemed to be faltering. No bulletins on the subject of pest control had been published since 1887.[65] Research during the decade bounced from subject to subject in an apparent search for some consistent program, including bulletins on "The Use of Gasses against Scale Insects" (F. W. Morse, 1887), "The Use of Hydrocyanic Acid against Scale Insects" ([Hilgard?], 1887), and "Spray and Band Treatments for the Coddling-Moth [*sic*]" (Hilgard, 1887).[66] For five frantic years, with new species entering the state all the time, nothing about insect pests came out of the experiment station. The success of the vedalia suggested progress in one direction, but the university offered no project of its own. Then Woodworth published three bulletins between 1892 and 1899, when bug-related research took off. The station furnished insights into phylloxera, potato worm, citrus rot, grasshoppers, peach tree borer, peach worm, red spider of citrus trees, resistant vine stock, sulfur sprays for red spider, and arsenical insecticides—all produced between 1900 and 1905 and all under the direction of the chief economic entomologist. In total, out of eighty-two bulletins issued between 1899 and early 1911, forty dealt with insects or fungi and their elimination.[67]

Groves across the state provided living laboratories for these experiments. Woodworth believed in the immediate commercial application of new discoveries, and he liked to pack up his laboratory and head for the field, there to set up his beakers and burners on site, there to cut the distance between innovation and application to almost nothing.[68] He found growers more than willing to pay for this service. Before the Smith-Lever Act established a national extension service in 1914, universities operated their own extension services from out of their experiment stations.[69] In California, scientists and growers discovered that on-site research generated valuable knowledge about pests and their habits, and they continued to work together in private experiments even after 1914.

When the peach growers of Placer County ran out of ideas about how to rid their land of the peach tree borer, they called on Woodworth. In this county, most of the peaches grown for the eastern market occupied an area four or five miles wide and ten miles long, following the tracks of the Southern Pacific Railroad. Even though the growers had used tested chemicals and sprayed correctly, the region reported losses of three hundred thousand dollars in a single year. The experiment station placed one of its assistant entomologists in the field, and the people of Placer paid all expenses for the investigation.[70] Results from these experiments and the newest methods of eradication came out in the local press. Losses fell from 50 percent to 1 percent in one year.[71] Other federations, public and private, also raised money to compensate scientists: the apple producers of the Pajaro Valley for help against codling moths, a similar group in Turlock to fight vine hoppers, and the orange growers of Los Angeles County to eradicate red spiders. The Southern Pacific and California Northwestern Railroads paid all transportation charges for Woodworth and his students, and the published reports on the results of these experiments ran out of stock.

With their duties completed the experts returned to the Berkeley campus, leaving behind prescriptions for chemicals that growers, henceforth, had to purchase in order to maintain their trees. While the university provided formulas and experiments, its staff could not mass-produce the chemicals themselves. Growers once exchanged recipes for idiosyncratic blends of the familiar soaps and oils—products in stock at their local supply store. But as the mixtures became more complicated, more dangerous, and more costly, some suggested that private companies take over responsibility for their production.

Arsenicals could not be made easily at home. Paris green required many different chemicals "which must be put together in the right proportions, and at just the right time," this according to George Colby, a chemist at the Station. Making Paris green required, besides experience, " 'striking tanks,' crushing, sifting, or bolting apparatus—all of which is quite beyond the reach of any one individual."[72] Colby, a colleague of Woodworth's, taught readers how to make arsenate of lead safely, but the process took time and mistakes could be costly. Growers were more likely to trust a commercially made product, especially if it had the endorsement of the experts in Berkeley. One grower called the new chemicals "impracticable unless put up already for sale by some one who is prepared to do so in the exact proportions."[73] Store-bought poisons would be cheaper, they argued, and would be backed by a guarantee from the company.

Thus growers became a willing market for orchard chemicals, creating a niche for a new industry. Woodworth smiled on the commercialization of spray chemicals: manufacturing and mass-distribution seemed like the next logical step in the forward march of California agriculture. Yet the increasing dependence of growers on packaged poisons raised questions of authority and regulation. The emergence of orchard sprays as a business meant that the university could no longer influence the quality of ingredients. When the local supply store offered simple materials for simple formulas, quality posed few problems. By 1911 prepared mixtures existed for most of the recipes in use at the time.[74] And when private companies competed for the pest control dollar they sometimes offered poorly prepared chemicals. When growers unwittingly used mis-represented and falsely advertised products on their trees they revealed incompetence and graft in the manufacture, distribution, and sale of in-secticides. To protect millions of dollars in fruit as well as its own repu-tation, the university moved to regulate the industry.

The companies that manufactured arsenical poisons tended to be those for whom arsenic was a by-product of some other process. Discov-ering a market for their waste, they packaged and sold it through the rural press. London purple came from the aniline dye industry. The first samples bore the name Hemingway and Company, an English manufac-turer of dyes that changed its name to Hemingway London Purple Com-pany when the new product caught on in America.[75] Gas lime, used against the woolly aphid, came from the refuse of gasworks. The sci-entist who tested it for the experiment station warned, "It is a strong medicine that must be used with care."[76] Sherwin-Williams Company of Cleveland, Ohio, makers of paint, developed their own lead arsenate. The directors of these companies had only the dimmest understanding of how farmers used their products.

Although the convenience of commercial insecticides made them ini-tially attractive, orchardists who allowed private enterprise to prepare their chemicals suffered unanticipated consequences. A bag of dry white powder called "No-Life," or "Kills-A-Bug," or "Conkey's Fly Knocker" came in the mail; the buyer prepared it according to limited directions, put it in the tank, and hoped for the best.[77] Growers often had no idea what actually came out of the nozzle, and no reliable test existed that might indicate whether or not the liquid would denude the trees or do nothing at all.[78] In 1900, the Bureau of Chemistry in Washington, D.C., tested forty-five samples of Paris green and found that only thirteen of them contained acceptable amounts of free white arsenic—71 percent were judged too dangerous to spray on foliage.[79] The *Pacific Rural Press* reported that "thousands of dollars have been lost by using Paris green which has been debased by adulterants. Cases have been known where Paris green has had less than half of its proper content of poison."[80] On the other extreme, some companies used free arsenic to adulterate their products as a means of increasing their weight and price. A chemist in a laboratory could wash out the excess arsenic or determine whether the sample would work against insects, but a fruit grower could not.

While no one was looking, insecticides had become a business perme-ated by fraud. American farmers consumed fifteen hundred to two thou-sand tons of Paris green in 1903, and the constant increase in demand encouraged manufacturers "to put upon the market products which have been carelessly or hastily manufactured."[81] Dishonest distributors contaminated the products they handled by removing some of the powder and replacing it with chalk or sand. In 1900, almost all Paris green came from New York, and Colby maintained that the stuff became more adulterated each time it exchanged hands on its way west.[82] Over half of the samples submitted by growers for analysis in 1903 indicated contamination.[83] The *Pacific Rural Press* published a letter from one confused consumer who asked, "Is there any way of ascertaining the quality of Paris green without a chemist?" In this case, the buyer did not suffer an overabundance of arsenic, but a lack of it. He counted on the product to save his pears from the codling moth but ended up with a crop of worm-eaten fruit, only 10 percent of it marketable.[84] Although the university could be confident about the veracity of its general recommendations for eliminating insects, it could not guarantee the quality of the powders and pastes available for sale. Fraudulent chemicals placed the university in a position where its suggestions had the potential to harm.

Only the state and the university working in conjunction could protect the most important industry on the Pacific Slope from the chemicals that had become essential to its product. C. W. Woodworth fused the power of both institutions into one document by writing the nation's first pesticide law in 1901. The "Paris green law" (soon to be called the California Insecticide Law) stipulated that all arsenical insecticides sold within the state be registered with the university, which would then issue a certificate of purity to distributors. The state then alerted the public that "no one should buy it [Paris green] without seeing this certificate" and levied a fifty dollar fine on any merchant caught violating the law.[85] The law set a legal standard of 4 percent free arsenic for Paris green—the same tolerance adopted by the Bureau of Chemistry.[86] Woodworth himself became the law's chief enforcing officer. Soon after the law passed in Sacramento, the apple growers of the Pajaro Valley sent a resolution to the state Board of Horticulture complaining about adulterated Paris green purchased from two San Francisco supply companies. The board informed the firms that, failing to show cause, each would be prosecuted by the state attorney general.[87] With this kind of authority on guard, the quality of orchard chemicals quickly improved.

The university soon protected the largest market for agricultural chemicals in the nation.[88] In 1911 California had 650 licensed dealers and manufacturers of insecticide. Pharmacies and small drug companies made the majority of retail sales. Establishments like Shulman's Pharmacy in

Los Angeles and Whitlock Drug Company in San Bernardino served as distributors for many different brands, including their own.[89] Woodworth reported that the state consumed close to $250,000 in insecticides annually, and he predicted a time when California would itself become a leading producer.[90]

Industry Leads to Excess

At the turn of the century a new pest control industry, one regulated by the strictures of law and science, came to its feet in California. With the passage of the California Insecticide Laws of 1901 and 1911 and the national Insecticide Act of 1910, companies doing business in California no longer resisted the university's standards—they embraced them.[91] The makers of tainted Paris green slowly realized that university authority pointed the way to a giant market share and that this authority, while it could not be stolen, could be borrowed liberally. Corporations emerged in the persona of conscientious scientists tending to the needs of the bug-besieged fruit business. They knew quite well that though regulation determined the purity of chemical products, no law regulated the extent of their use. The material itself had been judged safe for trees, but was it safe for consumers and the soil? The companies could not be trusted to speak on such issues, since they had no incentive to restrict the application of their products. When companies became the distributors of insecticides, they brought intentions different from those of the university. The formation of one company in particular demonstrates the process by which "scientific" insecticides became the basis for private enterprise.

The codling moth came to the Pajaro Valley, and nothing the growers there threw at it did anything more than slow its advance. In a single season, the insect Woodworth called "a pest of first importance" destroyed $500,000 in fruit in the largest apple-producing region in the state.[92] The orchard association of Santa Cruz and Monterey Counties requested the attention of the experts at Berkeley before the infamous insect wrecked the local economy. They raised $2,750 in 1903 and $500 more a year later to fund an entomologist, and in 1904 Woodworth dispatched the assistant superintendent for the University Extension in Agriculture, Warren T. Clark, to the town of Watsonville, ninety-five miles south of San Francisco.[93]

Clark recommended Paris green and initiated a spraying program, only to stop it abruptly. The atmospheric humidity of the Pajaro Valley caused the release of excessive amounts of arsenic acid, resulting in widespread damage.[94] The financial toll from several hundred acres of denuded trees almost equaled the rampage of the codling moth. Clark accepted another position and left the project. In his place came two of Woodworth's students, William Hunter Volck and Ellerslie Edgar Luther. Volck had been testing oil emulsions against the red spider in Southern California when Woodworth called him to Watsonville.[95] Luther joined him sometime between 1904 and 1906. They concluded that the damp air of the Pajaro Valley rendered all available commercial samples too volatile. In response, Volck and Luther invented a compound that bound the arsenic more tightly to the lead, lowering the solubility of the arsenic acid and slowing its release. The resulting product left undamaged even the most tender leaves, making it safe for greenhouse plants and household gardens. They called it "basic" lead arsenate.[96]

To the reader, such an invention might seem little more interesting than a wormy apple, but it fundamentally changed the insecticide industry. At a time when flimsy companies boasted secret formulas with bogus claims, two university-trained scientists, both fed up with the poor quality of existing insecticides, invented what was assuredly the very best poison of its kind and considered ways of making their discovery public. Presenting growers with a new formula by publishing it in an experiment station bulletin would not put their invention into many hands. Only a factory working under laboratory conditions could manufacture the spray. After consulting with Woodworth and after raising $5,000 from family members and interested investors in Watsonville, Luther and Volck founded California Spray-Chemical in 1907. It was the first insecticide-manufacturing company dedicated to "scientific pest control" and products guaranteed "Correct, Proper, Right, Straight and Pure."[97] California Spray-Chemical, selling under its ORTHO brand, made scientific research the engine of market innovation.

The popularity of Ortho's basic lead arsenate and the company's upright manner of doing business soon encouraged other companies to seek the same respectability. By 1913 companies all over the state had developed their own versions of Ortho's persona as well as their own versions of Basic Lead Arsenate. Some of these corporations included Bean Spray Company of San Jose, known mostly for their spray equipment; Braun Corporation of Los Angeles, makers of fumigation tents and cyanide gas, used mostly for orange trees; Fruit Growers' Supply Company, a branch of the California Fruit Growers' Exchange that manufactured insecticides and fruit boxes; and Sherwin-Williams, which turned out arsenicals and other kinds of insecticides.[98]

The university welcomed these firms. In Woodworth's conception, private enterprise brought scientific solutions to agriculture in ways an experiment station could not. Far from posing a threat to scientific authority, private enterprise served the university's purpose by making laboratory innovation available to thousands. The extension service and the farm bureaus could propagate information, but they could not put surefire bug spray in the work sheds and mixing tanks of California's blooming population of fruit growers. Woodworth himself became an active participant in the research and development of private companies. In 1915 he accepted an invitation by the Braun Corporation to teach a weeklong seminar on the techniques of orchard fumigation at Pomona, in the heart of the orange belt. Braun donated all the machinery and the American Agricultural Chemical Corporation gave the spray materials.[99] Present were some of the major orange growers in Southern California, along with scientists who delivered papers and led discussions.

In meetings like this one, Woodworth encouraged private industry to do for orchardists what the university could not—disseminate invention. He believed, moreover, that a corporation conducting itself with the authority and rectitude of the experiment station performed a greater public service in its field than the station itself.[100] Addressing economic entomologists across the country, he emphasized the possibilities of private industry to do the insecticidal work of the land-grant colleges: "I am not sure we all appreciate the tremendous influence the manufacturers and dealers of insecticides are exerting. They are in touch with a hundred growers where an Experiment Station Entomologist reaches one. They have the last word when they furnish the goods just as they are about to be applied. Their advice will go far to confirm or to counteract our recommendations."[101] Would the manufacturers of insecticides— the people with the "last word"—accept responsibility along with authority? He had reason for concern.

Rather than provide a public service to growers facing financial devastation, the new corporations sought a share of their income. Toward this end, Sherwin-Williams understood exactly what image to assume. They said that consumers of lead arsenate had been "working in the dark," preparing their own insecticides at home. Instead, the company offered "materials ready-made which have been thoroughly tested and which are manufactured along the exact lines of scientific investigation," materials that had been tested, they continued, at "the various Experiment Stations."[102] Advertising departments propagated information they judged to be as objective and dependable as anything from a university laboratory, and sometimes it was. Ortho published a *Handbook of Economic Poisons* that detailed how best to mix and when to use their products. The company hired a team of servicemen and salesmen operating out of offices in six California cities to inform consumers about proper spray schedules, using information taken right out of station bulletins.

Not all companies acted as responsibly. Sherwin-Williams provides an example of how manufacturers ventured beyond the claim that a team of scientists had attested to the safety of their products. Far from posing a danger to foliage, the Sherwin-Williams "new process" arsenate of lead was said to actually enhance its qualities. "In fact, it is generally conceded by those who have used it to have a most beneficial effect on the foliage, tending to increase the rich, dark green, glossy appearance of the leaf and never causing any burning."[103] If this appetizing description did not increase sales, the company announced what it hoped would become a habit of mind among the public: "Spraying used to be looked upon as an expense; but it is now considered by all up-to-date growers and farmers rather as an investment," a necessary part of profitable agriculture. Where the university presented a set of conditions dictating when or whether to spray and what to use, many insecticide manufacturers told consumers that their products would get the job done in any situation, at any time of year, and with virtually no stated limit on how often to spray. In the era of the private company, the perils of insecticides did not come from their poor quality but from their perceived necessity.

Orchardists did not have to be persuaded to use pesticides. It paid to spray. Between 1900 and 1921 chemical protection for a mature orchard cost $10.00 to $20.00 per acre, depending on the number of trees per acre.[104] The way growers looked at it, they only had to spend $15.00 per acre to gross between $100.00 and $700.00.[105] Consider the cost of raising an acre of pears between 1918 and 1921. Annual orchard costs including pruning, spraying, taxes, and cultivating summed to $80.50 per acre. Harvesting consisted of picking and hauling for fresh and canned pears ($3.00 per ton) and additional labor charges for dried pears ($41.50 per ton). A mature pear tree yielded five tons per acre of fruit suitable for fresh sale and canning and another one ton fit only to be dried. So the cost per acre to cultivate and harvest a pear orchard can be figured at $137.00.[106] Subtracting this from a total gross profit per acre of $328.75 leaves $191.75.[107] Even on land valued at $1,000.00 per acre this net profit would have covered interest on the investment and still left a sum to be spent on the household and next year's crop. It required only $15.00 an acre to protect the orchard against possible ruin, a trivial cost given the potential profit. More than simply a prophylactic against insect damage, pest control emerged as another tool growers used to increase the profits of intensive cultivation. What Woodworth observed about the study of economic entomology in 1915 applied equally to the practice of orchard spraying: it had advanced from "methods of killing injurious insects" to "methods of making money by the control of insects."[108]

With their investments at stake, growers fell into habits of excess. Indeed, some found it impossible to use too much. Woodworth criticized the careless and impulsive manner in which many growers used insecticides, often beseeching his readers to apply them at the appropriate time and in measured quantities. Pesticides saved the state "hundreds of thousands of dollars" each year, he wrote, but "the farmer should clearly understand that the use of remedies is purely a matter of dollars and cents. Money or time should only be invested in this work when there is good prospect of an ample return." He told orchardists that treatments very often have no effect and result only in "a waste of time and money. . . . Fully half of what it now costs to treat our crops is wasted." He called preventative spraying as foolish as preventative medication, and yet growers did it all the time, subjecting the environment and consumers to dangerous chemicals.[109]

Woodworth did more than exhort. He defined for orchardists exactly when, why, and how they should apply arsenicals (and other products). With regard to the codling moth, he divided the apple and pear season into three periods or "campaigns" (keeping with the martial language of pest control). The spring campaign protected the blooming flower. Using platforms and long applicators, employees sprayed a poison mist over the tree tops to fill the blossom cups. This killed the mature moths before they laid their eggs.[110] The second campaign hit the hatching larvae. Woodworth instructed orchardists to spray thoroughly and more than once during this critical time. The experiment station advised them to use a mist fine enough to be carried off by the wind. Trees should never be sprayed to the point of dripping. The most important campaign— the third—stopped the young and hungry larvae before they stripped the trees and invaded the fruit. The war against the bugs had its own seasonal cadence, but not even the professor could say with certainty how often a patch of ground should be sprayed.[111]

Even with the best advice to guide them, growers often sprayed too much. Woodworth said that spraying and fumigation only paid off when the cost of materials and labor amounted to less than the damage. But the very purpose of pest control was to *minimize* damage—to push the boundary of losses from insects to as close to zero as possible. The *Pacific Rural Press* reflected the confusion over the proper use of chemical cures. The editors claimed that Paris green delivered 80 to 90 percent of the crop, while only 10 percent survived to harvest without it. Yet the same writer observed that orchardists often sprayed too much to achieve these results: "Failures have followed from improper spraying; using a shower, instead of a fine mist, and allowing the poison to flow off the foliage and fruit in the running or dripping water." [112] A grower could not easily determine how many times to spray since no one could say how much poison would eliminate all traces of an insect. It depended on the skill of the sprayer, the quality of ingredients, the exactness of application, and luck. To stop at three sprayings when the trees might need five would be to put the crop at risk. Better to spend more than to stop short and watch the season's investment get eaten up. To be judicious might be to use too little, and then the entire expense would go to waste.

Strange as it may seem, few people at the time questioned the practice of soaking the nation's fruit supply in a solution of lead and arsenic. Though common sense might have led them to question the practice, most growers saw no indications that spray chemicals might be dangerous—no mass poisonings, no toxified orchards. Then in 1919 something happened in a far-off city that temporarily focused attention on the insecticidal habits of California growers.

In August of 1919 an inspector for the Boston City Health Department accidentally came upon a shipment of pears at a fruit stand that had been coated with an unidentified white powder. He took the pears to the lab and found high levels of arsenic and lead. The health department quickly surmised that the pears had come from California and ordered all such tainted fruit to be removed from displays throughout the city. [113] The department then contacted the Bureau of Chemistry, the agency given the authority under the Food and Drug Act of 1906 to enforce interstate commerce with regard to adulterated food. The city's next step was to call a meeting of health officials in Washington, D.C., where representatives from Boston informed officers of the Agriculture Department that America's food supply was being contaminated. [114]

The pear growers took none of these complaints seriously. Frank T. Swett, founder of the California Pear Growers' Association, called the confiscation of his product in Boston a "wholesale and ill-advised condemnation of pears."[115] He clearly tried to disperse any sense of panic among the members by announcing that "a rational solution of the problem by sensible action at both ends of the line will ensue, and the results will be worth hundreds of thousands of dollars to our growers." Whatever that meant, he gave no indication that standard practice would change. The orchard owners insisted that they ran only a few steps ahead of the codling moth and that they would not relent in an essential practice just because Bostonians had never seen lead arsenate on a pear. The Sacramento River District lost over $150,000 to worms in 1919, equivalent to a "tax" on every orchard of $75 to $150 per acre. Statewide losses approached $500,000.

Swett answered the controversy by spraying double the amount of insecticide—an amount roughly equal to six sprayings a year. Rebuffing the officials in Boston, Swett issued a bulletin to the association in the spring of 1920 in which he asserted his confidence in chemicals and encouraged members to go out there and make money by the control of insects: "Brethren, let us spray. . . . Unless we give the job 100 percent personal attention, what is the probable condition today? Perhaps 80 percent of the pears are coated with arsenate of lead. Then the other 20 percent will be the happiest possible homes for worms, robbing you of $100 per acre. . . . If in doubt, why not repeat the first spray, immediately, and make sure?"[116] Swett proclaimed a dubious victory: "For the first time in history, wormy pears were conspicuous by their absence. The campaign saved the growers $200,000 and the consuming public were likewise benefited."

Technological implements reminded Americans that they had become reliant on methods and means of food production that they knew nothing about. Insecticides kept the shine on the apple, but they also raised the question of the public's health.[117] By the late 1920s lead arsenate had become a feature in the cultivation of Florida oranges, Mississippi cotton, and truck garden vegetables from New Jersey to Colorado, and scientists feared the effects of long-term and continuous exposure to poisons and toxic metals on the well-being of unknowing consumers.

Doctors began to present evidence of the effects of small amounts of lead and arsenic on food products in the late 1920s and early 1930s. A team of doctors reported that "clouds of poisonous sprays" fell each year upon trees, shrubs, vegetables, and standing water, and they said that "this warfare is being waged in every garden, orchard, and cotton field throughout America."[118] These same doctors concluded that small amounts of arsenic and lead in the diet might cause "eczema, keratosis, peripheral neuritis, disturbances of vision, and neurological symptoms hitherto obscure." [119] One cryptic note in the *California Cultivator* from 1909 tells of a young boy from Hollister who "ate cherries and drank milk and died within an hour." [120] Researchers examined hundreds of samples of fresh and prepared foods with mixed results, but they agreed that the use of lead arsenate constituted a threat, one all the more insidious because it caused no emergency and prevented the public from taking note.

Unlike other poisons, spray chemicals usually cause no obvious sickness or immediate death in the afflicted. One investigator at the USDA commented on the "prevailing misapprehension" among the consuming public "that because they themselves have never been seriously injured by contact with spray materials or by eating sprayed vegetables, there is no danger to human health." The officials did not worry about people falling over in the streets after buying pears from a pushcart but about "those more remote and insidious chronic effects which follow the frequent administration of small amounts of poison over a period of several years." [121] Although they could establish the toxicity of the chemical in a laboratory, following the trail of lead arsenate into the consuming public to discover genuine cases of poisoning continued to be difficult. Only in rare instances could doctors trace illness and death to a source. [122]

Not surprisingly, California growers made no comment on the subject, not wanting to attract any attention to themselves. They only mentioned it in defense and ridicule. One sneered that he doubted "whether a peck of fruit containing 1½ parts in a million of spray residue has the tummy ache potentials of a single under-ripe unsprayed apple or pear." The only people in any danger were the "harassed growers," whose troubles from health officials and the press "are undoubtedly temporary." [123]

Neither did orchard owners give much reflection to the possible damage their trusted chemicals inflicted on the soil. Levels of arsenical residues in unsprayed soils averaged 2.5 to 10 parts per million (ppm) but reached levels as high as 335 ppm in soils subjected to years of spray-chemical application. [124] Reports of mysterious damage to trees across the irrigated West prompted a handful of scientists to consider how the land fared under the rule of the spray gun. Beginning in 1910 a controversy broke out in the pages of the *Journal of Economic Entomology* over the environmental consequences of arsenical chemicals. The discussion began with Bulletin 131 of the Colorado Agricultural Experiment Station, in which the author, William P. Haedden, alleged that insoluble arsenic transformed into its soluble (and thus highly dangerous) form after it sank into the soil. [125] The agent that changed arsenic from benign to hazardous was none other than alkali salt—a common element in many desert soils. "The insoluble arsenical compounds," Haedden maintained, "are being converted into soluble ones in the soil."[126]

The bulletin provoked an immediate response from proponents of spraying, forcing Haedden to publish a defense in which he repeated many of his original claims while attacking the remarks of his principal skeptic, E. D. Ball of Logan, Utah.[127] Yet although they disagreed on minor points and on exactly how the chemical functioned in the soil, Ball shared Haedden's central premise that alkali and arsenic made a bad combination. The next year Ball published a more detailed study of spraying in Utah in which he concluded that "where alkali is present in any quantity it is probable that the arsenic of the spraying solution will be set free and will assist in the injury of the trees."[128] Later research suggested that alkali itself did not determine the fate of exposed soils, that sooner or later soluble arsenic appeared in all fields and orchards that had ever been subjected to arsenate of lead and resulted in stunted growth or death to field crops planted in old orchards.[129] A concentration of 3 ppm killed alfalfa and barley in one experiment, and the concentration in many apple orchards could be one hundred times as high.[130] By the 1940s the amount of lead arsenate being sprayed in the state of Washington peaked at 56 kilograms (or 123.2 pounds) of elemental arsenic per hectare (or 2.47 acres). Farmers in some places used arsenic as a soil sterilant—literally to kill all living organisms—in a terribly misguided method of weed control. The amounts necessary for such practices ranged from 504 to 2,520 kilograms (5,544 pounds!) per hectare.[131] Orchard soil sprayed with considerably less lead arsenate showed signs of it for years after.

In spite of these dangers, and largely ignorant of them, growers used spray chemicals to spread intensive fruit cultivation all across the state. In 1900 the estimated worth of California's 28,828,000 acres of farmland was $630 million. Twenty-five years later, farm acreage had decreased to 27,500,000 acres, but total value had risen to $2.8 billion, a percentage gain of 344 percent.[132] In the ten years between 1899 and 1909 the production of deciduous fruits expanded from 22,690,696 bushels to 31,501,507, a 28 percent rate of growth. Grape production grew 57 percent in the same period. The number of bearing and nonbearing fruit trees dropped between 1910 and 1920, but the 1920s made up for the loss with a dramatic expansion in acreage. The state experienced a 63 percent rise in the number of farms, from 72,255 in 1900 to 117,993 in 1920. Insecticides were not solely responsible for this growth, but it would never have happened without them.

For those who depended on orchard sprays it seemed that scientists and manufacturers could do no wrong. In 1926 the editors of the *Pacific Rural Press* celebrated the relationship between growers and manufacturers with an entire issue dedicated to "The Battle with Bugs." The editors sang the praises of "the scientists behind the gun, likewise the ranchers who hold the nozzles that make our 'varmints' run. Probably in all this far-flung world there is no better sprayed area than California. And probably no business of any kind anywhere has any practice more progressive, better planned, or more conscientiously carried out than the spraying of California's premium fruits and vegetables."[133] Nowhere could one find an industry, wrote the *Press*, in which so many scientists worked for the good of the producers and so many producers put into practice the knowledge of scientists. And should another bug penetrate the borders and begin to lower profits and raise costs, "immediately the scientist gets busy. He scours the world for information about the new pest, photographs him, thumb-prints him, studies his likes and dislikes, observes his gluttony . . . and devises a poison cup for his undoing." Evolution also operated under the scrutinizing eye of the entomologist: "The 'bug' becomes immune to the standard remedy. . . . But the scientist is still on his heels, and the ha ha of the hardy bug is smothered in the new oil spray." No problem lacked a solution. Controlling insects was like beating back a persistent but inferior invading army for ends that (they imagined) Jefferson himself would have approved: "Eternal vigilance is the price of peace and of good fruits and vegetables."

That vigilance was tested yet again in 1929 when the most destructive pest ever to enter an orchard established residence in California. On a trip to New York in 1926 George Hecke examined a crate of grapes from Argentina covered with maggots—squirming larvae that he feared represented the first wave of the Mediterranean fruit fly in North America.[134] Three years later, Hecke called the bug "the most serious and destructive of fruit pests."[135] Congress made $4,250,000 available for its eradication, and the state of Florida, where the fly first appeared, gave an additional $300,000. This smallish, elegant fly caused so much fear that the senior entomologist for California and seven county horticultural agents went to Florida to learn what they could do about it before the fly arrived. They did not overestimate the emergency: in Hawaii the fly virtually destroyed the deciduous and citrus fruit industry. All efforts at eradication failed, including introduced predators. One photograph shows a citrus grove, its ripe oranges scattered on the ground with from ten to sixty maggots incubating in each.[136] In spite of all attempts to hold the border through the usual inspection and confiscation, the Mediterranean fruit fly hit the state and stimulated the chemical industry to develop new products.

With each successive winged threat to the orchards, the chemicals needed to fight them became more powerful. A speaker at one Farmers' Institute meeting reminded his audience that because of the ever-increasing variety of orchard pests "there is imperative need that the orchardist keep always in close touch with the latest and most success-ful methods of combating their ravages." [137] Industry answered with a stream of new products to make pest control cheaper and more effec-tive. From Ortho came "Volck," an oil spray said to be easy on foliage and death to the red spider and all kinds of pests. The companies them-selves changed, spent more money, grew larger. In 1930, California Spray-Chemical became part of Standard Oil of California, now known as the Chevron Corporation. The Ortho brand became associated with one of the largest industrial corporations in the world, the makers of hundreds of chemical and petroleum products. Chevron established factories in South Africa, Australia, and Italy and sold their products to farmers the world over. [138] The Standard Oil buyout represents the ris-ing importance of the insecticide market in California and the United States, and it predicted a time when all insecticides would be manufac-tured by large chemical companies.

Bug spray and big business came to the orchards to keep insects from turning rural California into marginal land. Insect-killing chemicals be-came a critical input that contributed to the capital intensity of fruit growing and maintained the state's relative advantages over other re-gions. Left unchallenged, insects lowered quantity and quality, in-creased the cost per unit, and ruined sales. Advantages of soil and cli-mate amounted to nothing when they ruled. Only record harvests and continued demand kept growers in business, and insects threatened both. California became the largest user of insecticides among the states because its delicate products were more vulnerable to the damage that insects brought. The university, state, and corporate edifices secured spe-cialized agriculture against its own tendencies to ecological simplifica-tion and helped to sustain a landscape of orchards to the horizon with-out a living insect in sight.

5

White Men
and Cheap Labor

Intensive, specialized horticulture required a labor system all its own. The work of the orchard, including pruning, irrigating, picking, and packing, could not be mechanized. One of the ironies of this most capital intensive of all agricultural industries is that it did not benefit from the reapers, tractors, and combines that so changed farm production beginning in the nineteenth century. Fruit growing remained a labor-hungry business in the midst of an automotive revolution. Insecticides ensured the size of crops and cooperatives protected their quality, but only a large and mobile labor force could move fruit from tree to packinghouse. This export garden—this mass-producing countryside— needed the work of hands. But rather than adjust the size of their orchards to reflect the limited number of workers available from family members and the local population, growers set out to secure enough people to produce fruit on whatever scale they found most profitable. Whether they owned twenty acres or two thousand, rural entrepreneurs refused to view labor as a limiting factor.[1]

California fruit growers depended on a floating population of uprooted men, women, and children to perform the tiresome tasks of horticulture.[2] Long before irrigated colonies appeared on the plains of the San Joaquin, certain farmers believed that agriculture in California might not survive—and would never prosper—without the people who visited rural districts at harvest time, worked for two to four weeks, and then disappeared down the hot valley highway until the time came around again. Fruit growers sought a commodity they called "cheap labor," composed of people who worked in heat, slept in filth, took what wages were given, cost next to nothing to keep, and returned the next year for the same treatment.

"Cheap labor" described workers who for reasons of poverty, discrimination, and displacement suffered the roaming life of the fruit tramp. The term does not define any particular racial or ethnic group, though race often determined who fell into it, nor is it a category to which we can apply a firm definition. Indeed, it immediately raises questions of consistency. Did poor and desperate people cease to be "cheap" when they went on strike or rebelled or otherwise demonstrated that they refused to be manipulated? Employers fashioned their own definition by the 1920s: cheap labor did the growers' work for the growers' wages and never negotiated from a position of strength. In fact, however, workers often negotiated with employers and won concessions. Yet growers at the top of the industry spoke an absolute language, one that stressed the necessity of pliable pickers and low wages, one sometimes intended to catch the attention of lawmakers. Cheap labor composed an input category— a necessary component of the specialized orchard—but like all such inputs it was never as simple or as predictable as the growers desired. Cheap labor emerged from the nature of the crops themselves as the only solution that growers could imagine for how to gather the harvest at a profit to themselves.

Yet for over fifty years members and supporters of the industry disputed each other and sometimes contradicted themselves about who should do the work and why. Indeed, this story begins with the growers cutting off their first dependable labor pool. Calling the Chinese a threat to their racially reclaimed civilization, farmers joined a vicious pogrom to banish them from the state on the premise that more preferable white workers would gladly take their places. But whites were scarce for a reason. No matter their poverty, they had a degree of mobility denied to people of other ethnicities. They could work in factories and lumber camps; they could choose within agriculture to avoid the worst drudgery. The controversy over race and the falling status of rural labor coincided with the gradual realization on the part of growers that specialized horticulture demanded a detestable kind of work.[3] Cheap labor evolved to the point that it finally defined anyone with no other options in the economy.

All of which brings us back to the nature of the crops. Just as it created the need for pesticides and cooperatives, so too did specialization create the need for workers. Though they found it much more difficult to control this aspect of the fruit business, growers did nothing to alter the pattern that the single crop traced across the landscape. Instead they pursued an equally monolithic vision of an industrial process over agriculture.

A Season's Labor All at Once

The fruit business needed many more laborers than did other types of agriculture, and it needed them to move from place to place. Work in the irrigated districts was seasonal. Periods of vigorous activity punctuated longer periods during which little work had to be done. Rather than accept hired men as members of their households for all or part of the year, growers favored a temporary arrangement in which word of a coming harvest brought people to earn wages for a few weeks.[4] What began as an informal and irregular relationship between roaming laborers and orchard owners became indispensable to the business. Wrote Richard Adams of the University of California's extension service, "Economic conditions have developed a well recognized floating population, mobile enough to move from section to section as crop demands require."[5]

Seasonal work had been the occupation of the disadvantaged and dispossessed before the advent of irrigated colony tracts. One account of Los Angeles in the 1850s tells of a "slave mart" in which former mission Indians were "sold" for a week at a time, "bought up by the vineyard men and others" to work the harvest. The author observed that the forced labor of Indians sustained an economy that quickly fell apart without them. After years of dislocation and violence fractured their numbers the Indians no longer served the vineyard men whose plantings withered along with the missions—"no longer profitable when cultivated with honestly compensated labor."[6] The wheat boom of the 1870s and 1880s also depended on itinerant men to cut, thresh, and sack grain. Wheat required little maintenance between the sowing and the harvest. The rootless souls known as "bindle stiffs" or "blanket men" worked in the fields for only eight weeks a year. With the gigantic crop in burlap sacks on its way to San Francisco, harvest workers traveled the roads and rails searching for income. Wheat and grapes made fortunes for their owners, but to migrants following the harvest, these crops predicted a life of unemployment and homelessness.

As irrigation extended across the coastal plains and interior valleys the need for seasonal labor spread with it, but rural progress was not supposed to look like this. Dreamers of California's rural future once claimed that irrigation would do away with the labor issue. Did New England have a poor and roving work force? Neither would the Far West once water had transformed its vast grasslands into a nation of small, diversified farms. William Smythe, the dean of irrigationists, said of the revolution to come: "With the supremacy of wheat will go the shanty and the 'hobo' laborer, to be followed in time by the Chinaman. In their place will come the home and the man who works for himself."[7] The application of water to land did foster the breakup of many large landholdings, and it created some of the smallest and most profitable farms in the United States, but the social designs of the irrigationists never came to pass. The climate of greed and gain nurtured a rural industry that used irrigation to bolster land values and profits, not yeoman independence. When conducted with the market and not a subsistence in mind, irrigated horticulture tended to specialization and required hired labor to a degree that made orchards and vineyards more akin to small manufacturing firms than to the humble farms in William Smythe's imagination.

Smythe did not understand the labor needs of even a small irrigated farm. Indeed, even the smallest fruit operation often proved too large for a family of four or five adults to tend. Consider the work of a twenty-acre raisin vineyard. Periods of close attention and tedious activity punctuated the seasonal cycle of grapevines. In January and February, after the leaves turned, the vines were ready for pruning. Workers carefully considered each plant to determine which canes (minor wooded shoots) should remain to become spurs (major branches stemming from the trunk).[8] Irrigation began in early March. With the rivers full of cold Sierra water, workers opened ditch gates to flood the land to a specified depth. Trees and vines "under the ditch" required constant observation. Vineyards near Fresno took between thirty and forty-two acre-inches of water per acre every growing season to produce grapes containing 85 percent water at harvest.[9] No other work compared in difficulty to the late summer harvest. Workers cut the berry sets with short knives and lay them on wooden trays to dry between the rows. The bunches sat for a week or ten days, after which they were turned by placing an empty tray on top and flipping the whole thing over. Another three to five days passed before the raisins could be taken to the packinghouse.[10]

For a family attempting to cultivate a twenty-acre vineyard the harvest presented a formidable amount of work. Five-year-old vines might bear as much as 5.5 tons of grapes per acre or about 100 tons on twenty acres (two hundred thousand pounds) after insect damage. Gustav Eisen, the local authority on vines, wrote that an adult man could cut between twenty-five and fifty twenty-pound trays of grapes in a day. This is not a very tight range, but to make the point that even a family of four superior workers would have failed to finish the job unaided, let us assume that each could cut the maximum—fifty trays or one thousand pounds of grapes in a day.[11] Four adults harvesting four thousand pounds every day could complete the harvest in fifty days. Then the same four people would have to turn every tray and pack over thirty tons of raisins into field boxes.

Market conditions and the fruit itself often called for faster picking. A slow harvest might result in stronger competition and lower prices. Then there was the question of perishability. Fruit left on the vine after it reached maturity ripened before transit and rotted before sale. If grapes, figs, and apricots for the drying yard were not placed in the sun at the right time they began to decompose. And if children rather than exceptionally strong adults did a portion of the work, our imaginary raisin harvest would have lasted even longer than fifty days. Finally, a point of critical importance to any family with a vineyard, the five-year-old vines considered here would yield close to 6 tons per acre the following year and more the year after that. Vines produced close to an additional ton per acre each year after the fourth—all the way up to 8 or 10 tons per acre for Thompson seedless grapes. Harvesting might have been a family-sized project on vineyards of five, ten, and even twenty acres with young vines, but with twenty acres in mature bearing, hired help became essential.

Crop specialization contradicted the commonsense wisdom of tillers in the older states. Before mechanized reapers, perpetual labor scarcity had one best solution: keep family members constantly employed. Husbands, wives, daughters, and sons all lived off of the productions and profits of the farm; they had to be fed, clothed, and comforted at all times. In order to pay their keep everyone from young children to grandparents worked at some profitable task each day. Farmers raised many different kinds of plants and animals, in part so that new seasons would bring new tasks, preventing a yearly time of idleness. Cows needed milking; chickens needed feeding. The prairie farmer might sow corn in the spring for a fall harvest and winter wheat for a spring crop. The animals had a market time and so did the fruits and vegetables of the kitchen garden. "The great art," wrote one authority in the 1830s, "is to divide it [labor] as equally as possible throughout the year." He added, "Thus, it would not answer in any situation to sow exclusively autumn crops, as wheat or rye; nor only spring crops, as oats or barley, for by so doing all the labor of seed time would come at once, and the same of harvest work, while the rest of the year there would be little to do on the farm. But by sowing a portion of each of these and other crops, the labour both of seed-time and harvest is divided, and rendered easier, and is more likely to be done well and in season."[12] In this way, cultivation on the diversified farm sought a balance between mouths and hands.

Farmers back East also knew that specialization meant paying out wages—a potentially damaging expense. The historian Clarence Danhof writes that farmers cursed the leeching of cash from their pockets in the form of payments to hired men. Many complained that "wage payments absorbed most or all of their profits and frequently most of the money income of their enterprise."[13] G. F. Warren, instructor in farm management at Cornell University, had more to say on the subject. Warren cautioned his readers to avoid the latest fashion in cultivation, including irrigated fruit. Sounding something like his colleague Liberty Hyde Bailey, he explained that people who entered agriculture to get "back to the land" or to make a fast buck often depended on the labor of others. "Very naturally," he wrote, "they compare farming with city work, but such a comparison is very misleading. They are usually attracted by the idea of extreme specialization and are likely to prefer some fad rather than a staple product."[14] Warren forewarned that labor is "in nearly all cases the most important item in cost of production." Wages drained away the yearly operating capital that paid for things like seed and repairs and rendered farming a flimsy business.

The California fruit orchard turned the labor economy of the family farm on its head. Growers did not make decisions about what to plant based on the available labor of family members. The warnings and cautionary tales of traditional farmers would not have dissuaded them from planting crops that needed a year's labor all at once. They found reasoned authority for their methods in the United States Department of Agriculture. A writer in the department's *Yearbook* acknowledged that although the object of good farm management was always to employ labor profitably in all seasons, "there are certain conditions which justify the farmer in disregarding the distribution of labor in planning a system of management."[15] "In certain localities it happens that a particular crop is enormously more profitable than any other," he continued, as when "a particular soil type produces a vastly better quality of product than other soils. . . . In such cases it may be advantageous to devote the land exclusively to a single crop. . . . Under such circumstances the high profit from the crop may enable the farmer to pay wages that will secure labor when he needs it."[16] Where a particular crop grew to famous quality it made no sense, in business terms, to plant a diversified farm. This logic narrowed the possible sources of labor. Because family work proved inadequate to meet the needs of even a small vineyard, and because it would have been too expensive for an orchard owner to accept into his or her household—to feed and clothe year-round—all the workers necessary for the short summer harvest, growers took their labor from the highways.

The Indians and bindle stiffs did not disappear with the end of mission farming and the decline of bonanza wheat. Contributing to their numbers were thousands of Chinese men who came to California to look for gold in the Sierra. High taxes and predatory whites drove Chinese miners from the gold fields and into the Central Valley.[17] And after the completion of the transcontinental railroad in 1869, Chinese workers returned to California and entered the rural labor market. Thus just as developers planted the first irrigated colonies in the late 1870s, California had in its midst a large number of people who had only limited access to land and capital. The men and women who moved from city to countryside searching for income composed a fluctuating pool from which fruit growers could purchase the labor they needed when they needed it.[18] Orchard owners did not feel compelled to plant diverse crops in order to better use a limited supply of labor because they perceived an infinite number of people whom they did not need to feed or house or otherwise support. G. F. Warren wrote that when cultivating fruits and other crops that could not be mechanized, "the supply of pickers limits the area that can be grown."[19] But if there had ever been a relationship between farm size and family size in California, the specialized fruit orchard severed it.

The number of people needed by the emerging industry should have caused every investor in irrigated land to pause. In his 1891 address to the State Agricultural Society William H. Mills estimated the labor requirements of Leland Stanford's thirty-nine-hundred-acre vineyard, probably the largest in California at the time. According to Mills, Stanford employed 70 people all year, 135 for a few weeks, and 700 for the harvest. Mills noted that the maximum demand equaled ten times the minimum. Extending these figures to the two hundred thousand acres of bearing vineyard statewide, he reckoned that 3,500 people could be continually employed, implying a total of 35,000 for the duration of the harvest. At peak season the state's vineyards required a land-to-person ratio of one worker for every six acres. Mills reflected, "It would be impossible to have the labor of nine men available for a few months in the vintage season for one man who might find steady employment," but this is exactly what the growers wanted.

Another way of estimating the labor demands of specific crops is to calculate labor days from harvest to harvest. Frank Adams, an economist with the University of California, estimated that it took a total of 254 days of work to cultivate and harvest ten acres of peaches. One person working full time with the proper equipment could supply 100 days of this work; the other 154 days had to be hired out. Apricots, for comparison, needed 208 days, and Adams recorded that one person was capable of furnishing 90 of those days and that a spouse and two children could make up the rest.[20] Labor is adequate in these examples, but they are estimated for only *ten acres*. An adult could do only 100 days of work in a peach orchard no matter what its size, but with forty acres the total number of labor days might jump to 400. Estimates for the total number of workers needed to harvest the state's fruit crop varied with the year and the time of year and could never be exact. By the 1920s observers suggested 48,000 people in March and 198,000 in September.[21] And the number of people needed increased every year as new land went into cultivation and as three- and four-year-old trees came into full bearing.

California became a place of very large and very small farms, but no matter how many acres they enclosed all orchards operated in much the same way.[22] Large-acreage orchards had few advantages over smaller ones when it came to labor, because the fundamental operations of fruit growing could not be mechanized. According to Lloyd Fisher, an economist and author of an important study on harvest labor in California, specialization made for "large scale" methods even on acreage that farmers in other regions would have considered tiny. Regarding crops like table grapes, lemons, oranges, apricots, figs, and peaches, Fisher concludes: "These are crops which cannot be handled by family labor, either with or without a hired man. Even on the smallest farms the labor requirements of the harvest far exceed the family supply available. Ten acres of peaches, tomatoes, or apricots cannot be harvested without a force of seasonal farm laborers. . . . Small farms are as dependent as large farms upon seasonal hired labor, resident owners as dependent as absentee corporate owners. The harvest labor market is primarily a function of specialization in labor-intensive crops, and the intense seasonal demand for labor which results."[23] Consider this remark from an economist with the USDA writing in the 1940s: "Generally speaking, the high seasonal labor requirements are a function of type of farm, and not of size. This is especially true on fruit farms." The definition of large-scale, in other words, cannot be pinned to the dimensions of the land under cultivation. Though small by comparison with farms that produced staple crops like wheat and corn, the twenty-acre vineyard was, in fact, very large if measured in terms of the labor and resources it commanded. Oliver Baker observed that an orchard of twenty-five acres might be worth twenty-five thousand dollars and require fifteen hired hands in the first weeks of fall. Baker had no doubt that a ten- or twenty-acre orange ranch in California may be as large, "measured in value," as a one hundred- to two hundred-acre farm in the corn belt: "The amount of labor required for operation is also often as large, and at picking time may be larger."[24] Though they differed in the complexity of their management, small and large orchards did not operate differently.

Dependence on workers kindled anxiety over the supply. Shortage provided a topic of discussion wherever growers met, and they often asserted that without plentiful workers their ability to produce fruit at a low cost was seriously impaired. A. B. Butler, a grower with over one thousand acres in vines who employed 225 men for harvesting and 450 men, women, and children in his packinghouse, expressed concern for the labor supply in 1890: "The labor question is to me a serious one. We must have plenty of hands during certain months of the year; during others again we need very few. Fresno alone will[,] in a few years from now[,] employ 12,000 in her vineyards and packing-houses."[25] One speaker at a convention made this statement in 1902: "Labor is the greatest factor of expense in any industrial enterprise, and this is particularly true of the fruit business of California. In order to successfully operate the orchards and vineyards there must be at hand an adequate supply of efficient labor."[26]

George Hecke noted on another occasion that the unavoidable irregularity of demand made the supply uncertain, "frequently resulting in disaster to crops and material loss to the producer."[27] The only alternative to an aggressive campaign to secure labor for the orchards, as Hecke saw it, was to abandon land in cultivation and return to "primitive methods in which the labor is not such an essential factor." By primitive he meant family labor and home consumption.[28] Hecke's attitude and the tendency among growers to ask for more labor than they needed to ensure their supply led labor economist Lloyd Fisher to conclude that in California, "the farmer's demand is more or less as he states it."[29]

The California Development Association, an organization dedicated to the state's commercial progress, detailed the chaos that would result if the tide of workers was either cut off by immigration restriction or depleted by competition. In general, stated their report on Mexican workers, "a sudden stoppage in California's sole supply of seasonal and migratory farm labor would constitute a dislocation in the agricultural industry which would be felt immediately throughout all industry."[30] They gave special attention to the needs of the small grower, noting that he or she too needed casual labor—"ten, twenty or fifty—and he needs that labor quickly to get his crop off and into market."[31] The report comes to a sudden point: "Fluid casual labor is this farmer's only salvation. It is the prime necessity of his success." In other words, there could be no fruit business without it.

The story of harvest labor takes its place next to the other subjects considered in this narrative. As with marketing and insects, so too with labor: growers demanded stability, used the power of the state to impose commercial order on the countryside, and insisted that they could conduct single-crop agriculture with the regularity of manufacturing. Yet the crops and the people who picked them never ceased to argue otherwise.

Confusion over Exclusion

When they talked about who should stoop for the harvest, employers spoke the language of preferred nationalities and despised peoples. Whether they hired Japanese families or Iowa farm boys, race became the criteria growers used to select their workers. Yet the crop system that implied a labor system had everything to do with this preference. The particular people they desired as labor depended on how the growers understood the nature of orchard work. In the 1880s Anglo-Americans had that work as their privilege; in the 1920s Mexicans had it as their misfortune. The story begins with the Chinese.

As developers laid out the first irrigated colonies, farmers and fruit growers participated in a movement to drive the Chinese from the state. After the Union Pacific Railroad connected the coasts, Chinese men moved into rural California looking for land and opportunity. Resentment on the part of white fruit growers, especially those with small holdings and modest means who competed with the Chinese for farmwork, along with the common perception that Chinese religion and culture could not be assimilated into a Christian society, finally reached the United States Congress, whose members moved to seal off California from these particular immigrants.[32] In a twisted agrarian rhetoric, supporters of the bill asserted that respectable work belonged to respectable white men, the same people who shed a virtuous sweat to build California into a white man's civilization. Chinese immigrants had no place in such a conception; indeed, they violated it by their presence.[33] The goal of the 1882 exclusion act, as expressed by John Enos, commissioner of the California Bureau of Labor Statistics, was to "give the labor of our State to the white people, thereby not limiting or decreasing their opportunities to work."[34]

But arguments to bar Chinese fruit workers met testimonials about their excellence as irrigators, pruners, and cultivators.[35] Some questioned the capacity of white men to do this work as well; others asked whether they would accept the mean toil and harsh living that newly arrived immigrants commonly suffered. Suggestions came from various corners that in order to convince white men to labor in fields and orchards, employers had to offer them more than the roadside for lodging. In the years that followed 1882, growers pursued a native-born, white work force and, at the same time, learned a few things about life on the migrant highway.

In his presidential address of 1883 before the California State Agricultural Society, Hugh LaRue called white people to the countryside to take up jobs still held by Chinese workers. The federal act of exclusion threw farmers and growers back upon their own resources. Orchardists facing the loss of their Chinese pickers and packers needed replacements who would give the same quality and quantity of work. LaRue broached the idea that this new source should come from white households: "Is there any apology needed if I admonish the farmers of California to rely upon and encourage their sons to become farm laborers?"[36] To fill the silence that likely followed these remarks, the speaker fell into a barren speech. Appealing to a fictional foundation of California, one without the complications of Indian peonage, LaRue strained to convince his audience that rural industry needed no immigrant labor: "Our forefathers planted here a mighty empire of population and wealth . . . without the aid of servile labor or the assistance of antagonistic races. They taught their sons that labor was honorable. . . . The question that confronts us to-day, shall our sons be taught like habits?"[37] LaRue spoke with gravity about the "experiment of exclusion," an experiment that "raises the great question of self-dependence, and the simple logic of that question, in the plainest language, is that our boys must be taught to work."[38] Yet LaRue's address also contained an important contradiction. When he endorsed the practice of planting only one crop—only the one that grew to perfection in specific districts—he failed to recognize that any agriculture that made "self-dependence" difficult to achieve made immigrant labor necessary.[39]

The members of the Agricultural Society must have considered that Chinese exclusion was indeed an experiment. If their sons refused to work, what then? No sooner had the anti-Chinese forces achieved their pernicious piece of legislation than questions emerged about the willingness of white men to take up farmwork. Whites performed all kinds of agricultural tasks in California, but they preferred to harvest wheat over fruits and vegetables because the latter required hand work, which they associated with the Chinese, while the former placed them in the more dignified position of teamster. Nor would white workers take the same wages as immigrant workers. J. P. Johnson, the owner of a San Francisco employment bureau, stated outright that white men would not work for $1.25 a day even with board, and that the orchard owners of Vacaville and St. Helena acted "in . . . their own interest" when they continued to hire Chinese workers.[40] The Chinese set standards that growers attempted to replicate with other workers, often without success.

Already by 1886 there seemed to be a misunderstanding between growers and the white men they hoped to employ. Some talked of a transition period and of necessary changes in the structure of farmwork. Commissioner Enos, who called Chinese people "a vice" and "a degenerating influence," admitted nonetheless that if they left right away "the farmers, orchardists, and vineyardists" would suffer "great loss and hardship" as a result.[41] W. D. Ewer, also the owner of an employment bureau, said the same. While he believed that white labor could take the place of the Chinese in time, he offered this observation: White men did not work like the Chinese. They demanded better housing, better food, and higher pay. Keeping them employed, he said, meant that "a change would be . . . necessary."[42]

The scarcity and expense of white labor became evident in the years after the exclusion act. Growers used Anglo newspapers to call women and children from across the state to come for a few weeks of picking: "100 men, women, boys and girls wanted in Maywood Colony to pick and cut fruit. . . . We want you now. Come quick! Bring camp outfit, or board in private family or at hotel."[43] The difficulty never passed. By 1902, growers reported a 25 to 50 percent rise in wages paid.[44] Grain and hay farmers, hop growers, and dairy owners also reported paying more money for fewer workers. Over the next five years railroad construction and industrial jobs offered better pay than agriculture and siphoned off part of its work force. George Hecke stated that "intelligent and reliable white labor is no longer content to engage in menial occupations" now that urban industry had better jobs to offer.[45]

Determined to create a white labor pool, and not comprehending the many reasons why whites did not remain in farmwork, the members of the 1902 Fruit Growers' Convention formed a committee, which included Frank Swett and George Hecke, to solicit men from the Middle West to move to California.[46] The concept came from H. P. Stabler, a fruit grower who argued that "an intelligent, thrifty, energetic, steady young white man who was raised on a farm can do more work than any laborer a fruit-grower can secure."[47] By offering them "a new country with a mild climate," good wages, and an "interesting business," Stabler felt sure that the sons of Illinois and Iowa would come by the thousands.[48] The California Employment Committee hired speakers to travel east and talk up the state. Canvassers visited towns in Kansas, Nebraska, Missouri, Iowa, Illinois, Ohio, Michigan, Pennsylvania, and several southern states, "delivering lectures illustrated with stereopticon views of California orchard and farm scenery" and distributing "a vast amount of literature."[49]

By the time of the 1904 convention—two years after Stabler first proposed the plan—the committee had succeeded in attracting about one thousand people, a figure that fell below expectations. The growers had underestimated labor shortages in other states and the desire on the part of young people to move into commercial occupations rather than cross the continent on the shallow promise of two dollars a day in a vineyard.[50] The publication of their booster tract, "Grasp This, Your Opportunity," of which they printed and distributed one hundred thousand copies, was more likely to attract tired farmers to suburban developments. The Iowans who came to California in the early twentieth century preferred the offshore breezes of Long Beach to the furnacelike heat and unrewarding toil of Fresno.

The resettlement plan failed for reasons that few at the time wanted to admit. The scheme was born on the premise that the same young men who worked as family-housed hired hands would choose to move to California, but news of the life they were likely to find in the Far West traveled far in advance of the boosters. One grower, a Mr. Cunningham, told his fellows that once while still a farmer in Minnesota he employed a hand who had spent a year or so working his way through the fruit districts. The young man said "that he wanted no more of California, that they wanted him to take his blanket on his shoulder and go into the stable to sleep." Cunningham reminded the convention that farmworkers in Minnesota lived indoors "and are treated as children of the house." Cunningham's hired hand expected the same hospitality from employers on the Pacific Coast, which was why he came home. Before leaving for California himself, Cunningham sent his own son to learn the fruit business and get settled, but after six months he too came home, telling his father "that he had not been treated much better than a dog by his employers."[51] Those who did make the move soon wanted to manage, rent, and own, sensing that farmwork in California carried none of the same dignity as it did where they had come from.[52]

Sleeping on the Roadside

The indignities of horticultural labor require a closer look. As orchard owners began to realize, the work itself selected the people who performed it. When white workers walked away from tasks they did not like, employers showed not the slightest willingness to accommodate them.[53] Increasingly, growers saw the difficulty of horticultural toil and the unenviable lives of the people who did it. They saw that harvesting demanded an unpleasant and sometimes dangerous existence. As early as 1902 one of them asserted that "there is not a [white] man who lives in any agricultural locality who wants to get in and do this work."[54]

The work subjected people to dust and insecticidal chemicals made from petroleum, sulfur, lead, and arsenic. It required work in trees where a fall from a ladder might result in broken arms, legs, or back. With piece-rate payment, the potential for heatstroke and the need to relieve oneself had to be balanced against the need to fill one more tray or pack one more box in order to make a living wage. Voicing an opinion that gained in popularity during the war years, one owner of field crops claimed: "You can not get white labor to go in the beet fields and on their hands and knees, chop out the plants; you can not get them to dig in the mud for celery; it is not the kind of work that white people engage in. You can not get them to raise the berries and the small fruits and go around on their hands and knees to pick them. That class of work can be done by coolie or some other kind of labor, and leave the more profitable lines of farming for our white people."[55] A despised kind of work required a despised kind of worker.

Accommodations were no more hospitable. Growers showed a reluctance to invest in permanent structures to house migrant help because buildings cost them money and took up space on valuable land.[56] They preferred to imagine the harvest as an opportunity for poor folks to camp out for a few weeks and sometimes claimed that workers enjoyed sleeping under the stars. But the lack of adequate sanitary facilities made this camping less than pleasant. A report on 876 camps inspected between April 10 and November 1, 1914—a period representing the picking season in virtually every crop district—reveals that out of 35 grape camps, 25, or 71 percent, rated "bad" for overcrowding and poor sanitation. Of the other fruit camps, 26 percent rated bad, 40 percent good, and 33.3 percent fair. Only 3 percent of all the grape camps surveyed had toilets at all, and of those 68.6 percent had "filthy" toilets. Filthy also described 36.6 percent of the toilets at all the fruit camps surveyed.[57]

Richard Adams and T. R. Kelly, both of the College of Agriculture's Extension Service, made a comprehensive report on labor during the war and revealed that the inadequacy of housing was "well known to all those acquainted with western ranch life"; but the two authors claimed, without evidence, that the sort of people who inhabited fruit camps did not mind being uncomfortable. Quarters provided for "peon, coolie, or Oriental labor" could be well below "American standards of living." The authors concluded that transient labor did not demand sanitation, a ready-made conclusion designed to justify urban slums at the edges of orchards and fields.[58] One grower admitted that he had seen ranches where "the horses were cared for better than the men."[59]

The flip side of life among orchards and fields was unemployment. While landowners contemplated which sea resort to visit for a few weeks or whether or not to buy their neighbor's forty acres, fruit workers loitered in cities and clung to highways. "Deciduous orchard labor in California may be classed as emergency work," wrote the *Pacific Rural Press*; "it is not continuous the year round, and the peak of the load comes during the summer months, and for the balance of the year labor has to look elsewhere for employment, which makes it a hardship, and especially is it the case with large families, as it is expensive for them to be on the continual move."[60] The management of C. C. Teague's Limoneira Ranch made a similar observation. In a description of that company's labor practices, a representative noted that although "modern intensive agriculture" shared many similarities with manufacturing, one important difference set it apart: output could not be regulated to keep a uniform number of employees busy all year. "Subject to the seasons, the vagaries of the wind, the rain and temperatures, agriculture knows its labor needs only from day to day."[61] The Limoneira Ranch, an operation with over seventeen hundred acres in lemons and walnuts, hired four hundred people at peak season, but kept only half as many regularly in a company-owned town.

The disparity between the number of people available to work the harvest and the number actually necessary for that purpose also became part of the fruit business. Growers had no idea how many people might collect in local labor markets or who would arrive in response to broadsides nailed to telephone poles across the state, yet they knew that broadsides called more people than they needed. When families arrived with their cooking equipment and bedrolls, employers often waited for the crowd to swell so that the wage they finally offered for a day's work would reflect a buyer's market. For example, in a report to Governor Hiram Johnson in 1914, the Commission on Immigration and Housing discovered that a Valencia orange district advertised an appeal for labor "long after it had all the men it could employ." In another instance from the same report, "an employment agent" operating statewide continued to call cotton pickers through the *San Francisco Examiner* for work in the Imperial Valley "long after the demand [had been] fully met."[62] In this way, employers took advantage of the unavoidable lack of communication among workers. They could call for people who had no information about competition, wages, facilities, or difficulty; they could turn away anyone rather than address legitimate complaints.[63]

Growers also used so-called employment agencies to secure their workers. These often fraudulent enterprises operated in conjunction with boardinghouses or bars or wherever transient people congregated. "The untrustworthiness of private agencies is now a matter of common knowledge," wrote the governor's commission. In their investigation of 81 out of the 247 agencies licensed by the state, 52, or 64 percent, seemed to be "of doubtful honesty." Agency owners admitted to a list of violations, including misrepresenting work, sending people to places where there was no work, false job advertising, splitting fees with foremen, and extortion from applicants.[64] A state law of 1913 regulated employment agencies and stipulated that applicants for jobs were to be informed about the place of work, wages, and sanitary conditions.

After regulating employment agencies, the state of California established its own, setting up offices in rural districts "from time to time" to facilitate the hiring of unemployed men. They opened offices in Brawley, Chico, Hollister, Marysville, Modesto, Lodi, and Newcastle—all

areas of intensive fruit and vegetable cultivation. State employment bureaus appeared at the request of local growers and civic organizations
who wanted migrants off the streets and out of neighborhoods where
they often begged for food.[65] The intention behind these temporary
offices was to make it "unnecessary for thousands of men and women
to roam from ranch to ranch looking for work." It hardly put an end to
the seasonal migration of workers from cities and towns to fruit regions;
instead, the labor offices signified that, henceforth, the state of California would have an interest in the agricultural labor supply.[66]

Wages added insult to injury. In 1917, a day's pay ranged like this:
general ranch work paid $2.00 to $2.50 per day, the same for picking
sugar beets. Fruit picking paid $2.25 to $3.00 for a day's work. These were
competitive rates compared with mining ($3.00 to $3.50 a day) and railroad construction ($2.25 a day).[67] California farmers (all crops) claimed
that they paid more in wages to their employees than farmers in any other
state by far—about as much as paid in Iowa, Missouri, Kansas, and Nebraska *combined*. Quite true: they paid out an average of $1,650 per farm
for board, lodging, and wages in 1919. But farmers in California spent
more on labor because they *used more* not because their workers lived
higher. Employers sometimes deducted board and lodging, so the cash
that landed in workers' pockets may have totaled less than the quoted
pay. Labor contractors also took their cut. Finally, no matter how much
money they received while working, harvesters faced unemployment
roughly between October and June (unless they could pick citrus from
January to March) and at odd times during the harvest season as they
traveled from job to job.

White pickers resisted and avoided these conditions. George Hecke
claimed that during the harvest season of 1907 "it was impossible to get
a sufficient number of reliable white men, *at any price*, to harvest the hay
and cereal crops and to gather the products of orchards and vineyards."[68]
And when those "good, prudent, industrious" white men—once called
the foundation of civilization—dared to speak their minds or declare
their independence from the knockabout life they led, they ceased to be
employable.[69] Growers complained about workers who challenged the
authority of managers, quit in the heat of the day, demanded good food
and housing, and in general demonstrated no fear of their employers.
One owner called them indolent, intractable, and unreliable. He offered
to pay white pickers $40 a month with room and board: "This is the
twelfth month in the year 1907, and I have there now the twelfth ranch
hand I have had this year."[70] Another grower once agreed to hire a crew

of white men to placate the anger of an anti-Chinese mob. Not all the men turned out on the appointed day, half showed up drunk, and most of the others refused to do the work. The employer set up a boarding-house with two cooks and bunks, but by the end of the first night only three men remained to finish the picking on a 320-acre vineyard. "How can you survive conditions like that?" the speaker pleaded. "We have come to the determination that if the whites will not harvest our crops we must go out and seek those who will."[71]

An important change took place in the industry's rhetoric regarding harvest work during the years leading up to World War I. Some called farmwork unworthy of white people, concluding that it demanded certain tasks that "white labor ought not to be asked to do."[72] Not a rung on the ladder leading to management and ownership, harvesting—especially the picking of field crops like strawberries and sugar beets—increasingly seemed degrading. One speaker addressing the Commonwealth Club of California declared that instead of workers earning their way to "independent proprietorship," the unwashed ranks of agriculture composed "a population of hereditary idlers and . . . a homeless, hopeless proletariat" where white people did not belong.[73]

White wage earners came to represent California's failure to absorb actual settlers. Thomas Forsyth Hunt, dean of the University of California's College of Agriculture, reflected in 1913 that a young (white) man in California could no longer purchase a farm without a banker to back the venture: "It is this lack of credit that forces [young men] into wage-earning occupations," he explained.[74] The same incremental land values that made so many fortunes in years past now spread the anxiety of diminished opportunity by forcing would-be freeholders to seek wages to support themselves. Harris Weinstock understood that the rise in real estate and the high cost of putting in an orchard spelled a booster's nightmare: "A frightfully large proportion of such investors have come to grief, have been forced back to the cities, many of them as unskilled laborers, to swell the ranks of the casual unemployed[,] and many of them have cursed the state as a delusion and a snare, have shouted their misfortunes from the housetops."[75] Nothing reveals the growers' disdain for farmwork as clearly as their discomfort with the thought that people like themselves might have to do it.

Harvesting had never been an honored occupation in California, but it finally attained all the prestige of peonage. "The truth is," said a member of the convention about work on his farm, "that only those seek it who are unfitted for other occupations or unable to secure more congenial employment." He called farmwork "the least inviting to our youth, the least followed from choice. Our farm help must be sought from some other source."[76] Sharing this dismal sentiment and convinced that the banishment of the Chinese had been a mistake, an opposing faction of the Fruit Growers' Convention that had grumbled since 1882 came out against the experiment of exclusion and the idea of white resettlement. The Chinese should not be prevented from immigration for being "servile," insisted members of this group; rather, they should be reestablished as California's very own homeless, hopeless proletariat.

"Our Work for Our Wages"

If the concept "cheap labor" had an originator it might have been John Powell Irish. Irish practiced journalism and politics in Iowa before leaving for California in the 1880s. He edited the *Oakland Times* and the *Alta California* until 1891, when he invested in agricultural lands. Irish's experience in the Iowa legislature, combined with his firebrand style of advocacy and his financial stake in irrigated crops, made him a palpable presence in the industry. John Irish is important to the present subject for his support of Chinese immigration in 1907 and for a similar defense of the Japanese a few years later. In both cases, he broadcast opinions seldom heard in public since the pogroms of the 1880s. Amid a labor shortage and frustration over white workers, he blasted the proponents of racial exclusion and outlined a new conception of the agricultural labor supply.

Irish maintained that the policy of exclusion rested on the erroneous assumption that the only people who should be admitted to the United States were those who could be assimilated into its white, Christian majority. He never questioned the belief that the unwashed castaways from strange nations made undesirable members of American society; he simply proposed that a national immigration policy could have an objective other than assimilation and membership. Irish asked the members of the convention to imagine a crop twice or three times the size of the one they harvested in the past year. Would they find enough reliable white labor? Remember, he told them, "the fruit on the tree and vine does not wait for a crew to go off to town Saturday night and get drunk and straggle back Monday and Tuesday, and some of them not at all."[77] Obviously white workers would not do; and after all, rural capitalists wanted more than just "cheapness of the labor," they wanted "fidelity and dependability," by which he meant obedience and docility. It was sometimes to the advantage of an enterprising people to admit certain others into their midst to serve in a restricted capacity but who, because of their obvious differences, could not be assimilated. The perfect laborers returned like the swallows year after year but never escaped their social isolation.

Irish pontificated with the Chinese in mind. Not only were they more capable workers than any of the people considered to replace them, but they posed no threat to the racial composition of American society. He called them "an orderly and law-abiding" people who would "do our work for our wages, . . . a most desirable form of labor."[78] Most of all, the return of the Chinese to the irrigated districts marked the end of a failed experiment in racial "self-dependence." Irish boomed to an applauding audience: "If the rural industries of California, the foundation of all prosperity, require this Asiatic non-competitive labor—a labor and form of immigration that we don't have to assimilate—we are not committing that hard task to our sons and daughters, to our grandsons and granddaughters, and our descendants."[79] Fourteen years after Hugh LaRue urged a worthy labor on the white sons of rural California, John Irish told growers that to secure the return of the Chinese would be to prevent their own families from falling into the drudgery and social indignity of farmwork.

The speaker only gave voice to an idea that many other growers had held for a long time. Immigrant labor could carry the burden of the white man's country without breaching its borders. But which immigrants would perform that labor? The period between 1900 and 1917 convinced growers that the workers they sought shared characteristics that cut across lines of race and national origin: desperation, uprootedness, and lack of organization.

Japanese immigrants did not fit the description. Though they arrived just in time to prevent a crisis, it became clear right away that these workers made capable negotiators who could not be easily manipulated. They stopped working at critical times, refused to scab against their own nationality, and pitted growers against each other to win higher wages. The Japanese engineered strikes to protest discrimination and unmet demands, and they boycotted employers who gave them trouble. In 1903 Japanese and Mexican workers in Ventura County formed an organization called the Japanese-Mexican Labor Association to negotiate contracts directly with growers rather than work through a contracting company owned by bankers and businessmen in the city of Oxnard. The organization held a successful monthlong strike and won major concessions.[80]

No workers who could force growers into a negotiated relationship qualified as cheap. Employers called the Japanese "cunning—even tricky" and a "pest in the shape of cheap labor," insults that only confirmed the success of these workers at playing the harvest game.[81] George Pierce, president of the California Almond Growers Association, concluded that they do not belong "either by training or instinct to the purely servant class" and that "their labor cannot be classed as cheap labor."[82] John Irish observed that the Japanese secured wages as high as three dollars a day, making them among "the highest paid farm labor in the United States."[83] All of this proved to the growers that the Japanese did not serve the purpose for which they were needed.

Yet when racial hatred came down on the Japanese out of bitterness for their success, Irish stood against it, calling the attack an "awful campaign of appalling falsehood."[84] After Theodore Roosevelt's "gentlemen's agreement" of 1906 halted their immigration, California enacted a series of Alien Land Acts in 1913, 1929, and 1930 to prevent Japanese immigrants from owning land. Laws that prevented capable people from investing their capital for the greater good of industry made no sense to Irish.[85] For the Japanese "the period of cheap labor was brief, and their wages rose" and rightly so, he believed, because the Japanese were not like other immigrants and any policy that limited their genius in any way also limited the future wealth of California.[86]

In other words, Irish advocated a brand of economic Darwinism, arguing that open competition among people of inherently differing capacities moved capital and led to economic progress. Though he considered the Japanese physically well suited to perform farm labor, he maintained that they should not be punished for escaping into ownership. The Japanese proved themselves against their Anglo-American employers on common ground—the market, an extension of nature itself, in his mind. Without judgment, without pity for the weak, the market selected for capitalist ability: "In economics there is neither hatred nor prejudice," Irish said. Indeed "economics" needed many contending peoples in order to thrive, exacting a "fine" from growers in the form of high wages as their punishment for excluding the Chinese. Yet although he meant it as a defense of the Japanese, this bland and vacant creed justified the broadsides and the employment agencies and the low wage with the claim that people rise and fall not because of restricted opportunity or manifest prejudice but according to their genetic fitness for the world. Those who ascend to ownership deserve to command the labor of others; those who sleep in the dust deserve nothing more.[87]

No one who did the picking saw it that way. On a hot Sunday after-noon, 3 August 1913, a rebellion at a place called Wheatland momentar-ily attracted attention to the labor practices of rural industry and sent progressives in the Hiram Johnson administration in search of measures to make migrant life more endurable. But the riot and the gestures of re-form that followed had little effect on the growers, most of whom re-acted more bitterly to the reform than they did to the riot.

When the Durst brothers plastered California, Oregon, and Nevada with broadsides calling workers to their hop ranch in the upper Sacra-mento Valley they attracted twenty-eight hundred people for far fewer positions. The Dursts wanted white pickers, but the crowed that gath-ered better represented the mixed population that combed the Pacific Slope for wages. The crowed camped out on a barren lot where nine toilets served everyone. Garbage, including animal entrails, lay uncov-ered and rotting. Water came from five wells, two of them dry and all of them distant from the fields where people worked in temperatures of over one hundred degrees Fahrenheit. As two hundred to three hundred children wilted, probably showing signs of heat stroke, the management sold ice cream and lemonade. Ralph Durst would not allow water to be distributed in the fields. Pickers willing to endure all this and a tent camp made of filthy rags learned that the Durst brothers offered a wage ten cents lower than the going rate. They did more than complain. They or-ganized, presented a list of grievances, and listened to speeches by mem-bers of the IWW. The Dursts panicked and summoned law enforcement. One of the officers fired a round into the air, and a battle broke out that left two pickers, a sheriff, and the district attorney of Yuba County dead.[88]

Wheatland signaled that labor relations in the irrigated districts had become combustible, that the furious disregard employers showed their workers could come back at them in kind. Nonetheless, growers ignored the riot. Indeed, the maneuvers of Japanese labor bosses caused dire pro-nouncements and urgent debate, but a deadly confrontation between police and thousands of workers caused no comment at the convention that year. Even though the local newspapers placed the blame squarely on the employer (three days later one reported: "As the situation begins to clear up it becomes more evident that the real cause of the trouble lies with the hop growers"),[89] the event evoked only a curt mention in the *Pacific Rural Press*. During the brief period of public scrutiny that followed Wheatland, revelations about the dismal lives of migrant work-ers competed for the public's attention with the investigation and sub-sequent trial of two members of the IWW. Richard "Blackie" Ford and

Herman Suhr, the two "agitators" accused of causing the riot, proved to be much more interesting to Californians than broken shanties and malnourished children.

Reformers strained to seize the moment. Under the leadership of Simon J. Lubin, Harris Weinstock's nephew, the California Commission on Immigration and Housing proposed upright and sanitary ranch camps as a way to avoid future uprisings. Yet although the act that served as its charter granted the Commission the power of subpoena and the right to hold hearings, Lubin could do little more than inspect and advise, without the force of law. A series of pamphlets issued by the commission provided growers with specifications for how to build roomy tents, tidy mess halls, and sanitary toilets. These plans for clean and ordered spaces, based on the theory that comfortable workers do not join the IWW, failed to persuade most employers to invest more money in their facilities. A number of growers believed the theory and cleaned up their camps, as though radicalism were a social disease that could be eradicated along with dysentery. Lubin, looking for some sign of success to put before the public, persuaded Ralph Durst to allow the Commission to improve his infamous camp. Durst claimed that the trial and conviction of Ford and Suhr vindicated his behavior and the quality of his camp, but with a possible IWW boycott of his ranch in 1914 he relented to reform. In general, however, the reformers and the growers spoke a different language. The latter had already decided that the people they hired deserved no better than what the Durst brothers provided at Wheatland. The few growers who called for higher standards to attract a "better class" of laborer received little attention.[90]

Growers finally paid attention when citizens attempted a ballot initiative for a universal eight-hour workday in 1914. In direct response, agriculture and its allies closed ranks and formed the Farmers' Protective League to protect their interests. Within a year, the organization claimed over one hundred thousand supporters and four hundred sympathetic newspapers.[91] Its officers and contributors included some of the most visible names in the state's rural industry: George Hecke, C. C. Teague, Harris Weinstock, G. Harold Powell, and Dean Thomas Forsyth Hunt of the College of Agriculture. George Hecke accepted the task of convincing voters to turn down the initiative when he wrote the argument for the opposition published in the 1914 election handbook. Hecke told the public that any statute that tied the hands of growers to make food as they saw fit hurt them on the international market and meant higher prices at the grocery store. The fair treatment of workers would also decrease the employment of white labor and increase leasing to Japanese fruit growers and farmers. In the ensuing labor shortage, white growers would be forced to "send wives and children into the field, as in Europe."[92] Hecke told the public that if they did not vote against the eight-hour workday, respectable people would join the peasantry.

The league and its allies defeated the measure but not the threat of reform. The issue had only just died when the leadership braced supporters for a new fight over a proposed minimum wage of $2.50 a day. "Do not make the mistake," cautioned the secretary, "that because we have won the eight-hour fight . . . our work is at an end."[93] Calling the opposition "fanatical," the league's president, F. B. McKevitt, told growers that they defeated the eight-hour law "only because we brought together the entire fruit industries [*sic*] and the business interests of the State."[94] When the growers cut their teeth on special-interest politics they realized that the Farmers' Protective League, or something like it, would be a fixture of the industry henceforth. In the coming years, they asked the public to look beyond obvious cruelties, to regard them as the necessary by-product of inexpensive fruits and vegetables.

Importing a Workforce

By the outbreak of World War I, growers commonly believed that Chinese exclusion had been a mistake, that the Japanese refused to serve as cheap labor, and that white workers would never fill the void in the workforce. When the United States joined the conflict in 1917 industrial employment opened up. In that year alone over one hundred thousand men on the Pacific Coast moved from agricultural work into occupations like ship building, munitions manufacturing, and lumber milling. These jobs offered higher wages than farmwork and shorter hours.[95] Strikes in Fresno in 1917, in which the IWW was said to be the instigator, also convinced growers that drastic action had become necessary. Faced with the unlikely prospect of attracting workers back to farming, and pinched by workers who understood the power they wielded over wages, employers supported a policy to import a labor force.[96]

To understand how fruit growers thought about the labor shortage, we have to know how they defined it. In this shortage no crops rotted on the ground, no fruit sat unpacked, no grower lost a year's income due to insufficient pickers. The most comprehensive report on the subject during the war years made it clear that shortage did not mean a lack of labor, but a lack of cheap labor. According to Richard Adams and T. R. Kelly of the California Extension Service, losses due to unharvested crops had been "surprisingly small" and "no state-wide losses in crops actually planted, or serious reduction in . . . cropped acreages, has resulted because of labor shortages."[97] Nonetheless, Adams and Kelly concluded, measured by rising wages and the difficulty of finding workers, "1917 offers abundant testimony of a farm labor shortage."[98] There could be no question in George Pierce's mind. The shortage "does exist," he pleaded to the Commonwealth Club of California, "and it will exist, and unless we get more labor, we are not going to produce the amount our Government demands."[99]

Agriculture scrambled for pickers, hiring anyone willing to work. A survey of 241 foremen taken in 1916 and 1917 found that 16 percent preferred white Americans, 26 percent preferred unspecified "foreign born" workers, and 58 percent had *no preference* regarding the nationality of the people who took wages from them.[100] Neither did they have a preference for age. Boy Scouts, members of YMCAs, and schoolchildren carried the burden when adult men and women could not be found. "Within two weeks of the closing of school," according to one orchardist, "a supply of over 2,000 boys reported to us as available." Employers negotiated with schools to delay the fall semester and to hasten the beginning of the summer break in order to make students available for the harvest. State officials considered child labor so important that California Senator Herbert C. Jones (a Republican representing San Jose) met with Dean Thomas Forsyth Hunt and S. Morris Cox, principal of Fremont High School in Oakland, to consider how best to employ children and how much to pay them. They agreed on $.25 an hour and an eight-hour day, also to include "proper housing conditions," "wholesome food," and the stipulation that no boy was to be sent more than one hundred miles from home.[101] White kids from Oakland were not to be exploited.

The war emergency brought others to help American farmers feed the free world, including the participants of one experiment called the Woman's Land Army.[102] A program of the National Council of Defense in cooperation with the United States Bureau of Labor and a product of the can-do patriotism that spread like influenza from every office building to every farm in the land, the Woman's Land Army was dedicated to bringing in the national crop while the men did their duty over seas. The company of women who went to California knew all about life on the migrant highway and understood exactly whom they replaced. Alice Prescott

Smith, the author of one article on the corps, gave her own interpretation of the farm labor contract in the West: "The grower looked over his goodly acres, calculated his crop, and went to his Chinese labor boss. 'John,' he would say, 'you find me fifty men. Come Thursday.'" When Thursday came, so did "fifty replicas of John . . . with mess kit and such bedding as they needed. They lived in the field, worked as the locusts, cleared the crop, and melted away." [103]

The women who joined Smith in California made themselves clear: They did not work like this. The Land Army presented a series of demands to the growers of Vacaville. They refused to work for a lower wage than men, nor would they work where men could be found. They would not bring in "outside girls" until the local population was fully employed. They demanded $.25 an hour for an eight-hour day and an increase later in the season to $.30 an hour, or $2.40 a day. (Recall that after Wheatland, growers organized to stop two things: the eight-hour day and a minimum wage of $2.50.) [104] They negotiated for a day of rest and for all camp buildings to conform to the regulations set by the California State Commission on Immigration and Housing. They would not, in other words, provide "cheap labor." The women set out to help with the war effort, not to be "cheapened by being exploited," nor to be "made an excuse for depressed wages or the supplanting of men."

Although growers attested to the ability and competence of the women, the program did not outlive the war as its supporters had hoped, nor did the standards the women enjoyed become regular practice. The growers of Vacaville paid good wages and bought tents and food in the pinch, but only until replacements could be found. As the Woman's Land Army worked in relative comfort, growers planned a more permanent solution to the labor drought, one that would not require schoolboys working close to home or regulation tents.

The memory of the Chinese had grown into a veritable Golden Age, and orchardists finally determined to replace them. They believed, however, that the need for tractable workers could not be filled with any group of people in the present population of the state. D. O. Lively of San Francisco argued before the 1917 Fruit Growers' Convention that the time had come to open the borders, at least for the duration of the war. [105] Perhaps the United States could allow the temporary immigration of men from the Caribbean Islands and Jamaica—the same people who had recently completed the Panama Canal. The proposed workers, according to Lively, "would fit the Sacramento and San Joaquin Valleys, . . . they are very law abiding people. . . . They flock very well, they stick together.

In that particular they are like the Chinese." Later in the same meeting, a vineyardist from Fresno delivered a paper in which he told about Chinese workers traveling to France by way of Vancouver and the Canadian railroad. He proposed that California growers take workers from abroad in the same manner—that they "import for the duration of the war a sufficiency of dependable farm laborers to do farm work only" and "send them home as soon as they were no longer required." [106] Here was a way to placate those who still insisted that California be sealed against immigration and, at the same time, ensure that growers secured workers who were newly arrived, politically powerless, and legally bound to their employers. When the soldiers came home they could take these jobs for themselves . . . if they wanted them. The convention of 1917 drafted a resolution to Congress: "Resolved that we ask our government as a war measure, to permit the introduction of sufficient labor." Later, the convention amended the resolution to read "Chinese or other farm labor" after a faction demanded that the Chinese be given preference in the request.[107]

In May of 1917 the United States Department of Labor answered the fruit growers of California, Arizona, and Texas with a ruling from the Office of the Secretary stating that, as a war measure and only for the duration of the war, Mexican nationals could be brought into the United States without a head tax and without passing a literacy test. The workers needed to furnish border officers with two pictures, fill out an identification card, and remain in agricultural work. Law and authority so utterly circumscribed their status in the United States that if these immigrants took employment outside of agriculture they could be arrested and deported.[108] This new legal caste also deprived them of the right to travel and live where they chose, and it gave growers unprecedented power.

The policy lasted longer than the war. A committee appointed by the secretary of labor reported that 50,800 Mexicans had (legally) entered Texas, Arizona, and California between the beginning of the program in 1917 and 1920.[109] A total of 17,187 had returned to Mexico, 10,691 had deserted their assigned work, and 22,637 were still employed in 1920. The committee did not demand that the Mexicans still employed in the United States in 1920 be sent back; to the contrary, they recommended that the Mexicans stay in their present occupations and concluded "beyond a reasonable doubt that white men are averse to accepting, and refuse to accept . . . employment as unskilled or common laborers." The government's survey disclosed that Mexican and Anglo-American laborers did not compete for jobs in agriculture to "any appreciable degree." [110] The policy had been "necessary and beneficial," according to the government, and with proper supervision at the border it could continue. By the 1920s the policy had resulted in a network of labor procurement that included chambers of commerce, state-owned employment bureaus, and labor contractors—all functioning with the approval of the United States Immigration Service.

Following the crops took migrant families on a rambling tour of the Southwest and California. The workers came up from Imperial, a bleached and searing valley with a few farm service towns, including Brawley, El Centro, and Calexico. An estimated 21,000 Mexicans, accounting for 35 percent of the population, lived in the valley in March of 1926, and many of them moved with the harvest. [111] Paul Taylor, a professor of labor economics at the University of California at Berkeley, documented the actual route taken by a Mexican family in 1934: Salt River Valley, Arizona, to harvest lettuce from January to March; Imperial Valley to tie carrots from March to June; Conejos to pick apricots in June; Tulare County to pick peaches in July and August; Fresno County to harvest plums in August; Tulare County to pick cotton from September to November; and then back to the Salt River Valley in Arizona to pick lettuce again until March. [112] The move from crop to crop only creates the illusion of constant employment. In fact, travel, waiting to be hired, competition for jobs, and the uneven and unpredictable duration of employment in each region prevented steady work. [113] The end of the picking season meant that children had to leave schools to travel north for crop peaks in the San Joaquin Valley. This constant interruption prevented many from ever finishing the first grade; they started over and over, even as they reached ten and eleven years of age. [114]

As they had in the past, growers appealed to an imaginary nature to vindicate the institution of migrant labor. Mexican workers appeared to be perfectly suited for a life of itinerancy, "a state to which the Mexican accommodates himself by nature much more readily than the white farm laborer. The Mexican is inherently a nomad." [115] One industrialist identified Mexicans as a "hereditary serving class" that would enable agriculture to be conducted "on a manufacturing basis." An infinite supply of supposedly docile labor suggested new possibilities for the business of farming. Now there could be great "dairy farms, milking cows by the thousands, orchards and vineyards numbering thousands of acres, for Big Business will find farming as favorable a field for exploitation as mining or manufacturing." [116] It all rested on the backs of workers. In 1928 the California Development Association concluded that Mexican labor supported the prosperity of California agriculture: "Restrictive immigration has shut out Europe and Asia from the California farmer. He has only the Mexican to turn to."[117]

California fruit growers had accepted the idea that cheap labor of any color was preferable to expensive and insubordinate white labor. Congress, however, had not arrived at the same conclusion. When the words "immigration restriction" came out of Washington, D.C., growers from the San Joaquin Valley impressed S. Parker Frisselle into service. Frisselle owned five thousand acres of vineyard land and served as chairman of the Agricultural Committee of the Fresno Chamber of Commerce. He appeared before the House Committee on Immigration and Housing in 1926 to stop the passage of the Box Bill (HR 6741), legislation that proposed to severely restrict Mexican immigration into the United States.[118] The bill's sponsor, John C. Box of Texas, opposed the importation of cheap labor on the grounds that the workers composed a class of peons and serfs that destabilized American society—an old saw among white Californians but one not in fashion at the time.

Congress was in no mood to hear arguments for the creation of an impoverished labor force, so Frisselle chose his words carefully, stressing the need for migrant labor and claiming (without evidence) that Mexicans did not remain north of the border but eventually returned home. The representatives confronted Frisselle with his own contradictions: If growers wanted labor so badly, why did they move to restrict it in the past? Why did Frisselle and his cohorts want to keep Chinese and Japanese out only to ask years later to allow Mexicans in? Commented one member of the committee: "Two years ago California came before this committee and stated herself in opposition to Chinese and Japanese immigration and in favor of Chinese and Japanese exclusion; stated that they wanted to develop a great big white State in California, a white man's country; and now you come before us and want unlimited Mexican immigration[.] . . . I can not see the consistency."[119] In fact, everyone present understood the consistency in Frisselle's position, though Frisselle himself refused to spell it out. "Let us be entirely frank about this matter," said representative William Perry Holiday, finally pinning Frisselle with the truth. "As a matter of fact, what you want is a class of labor that will be of sufficiently low type that they will not have the ambition or make any effort to become owners of any of your land. Is not that really what you want?" Frisselle had said as much in his description of Mexicans: "The Mexican is not aggressive, he is amenable to suggestions and does his work. He does not take the Chinese and Japanese attitude."

With the committee closing in, ready to deny the request, Frisselle abandoned argument and cut the matter to the marrow. "The problem for you to weigh," he said with blind force and tense finality, "is whether or not the development of agriculture is worth what it costs to carry on." In other words, if the gentlemen of the committee enjoyed their pears and raisins then they should allow western growers to make them by any means necessary. Cheap food carried a high price. Following John Irish, Frisselle believed that in order for growers to feed the cities they should not be restrained by an immigration policy that sought so-called desirable citizens.[120]

Continued immigration mattered more than ever, even as tens of thousands of Mexicans moved north each year, because the guest workers evaded their bondedness and did not stay in agriculture. "We lose them as rapidly as we do our own boys," said Frisselle. Keeping them on the job and out of the schools and employment offices turned into a policy of armed coercion. Where other workers could leave odious sanitation and low wages, the uprooted Mexicans were virtually forced into service and prevented from pursuing work as they desired it. Growers openly spoke about confinement and supervision as though the workers were prisoners in a chain gang. "As far as the Mexican labor is concerned," said the Santa Clara fruit grower Charles E. Warren, "if you take a Mexican out in the field, put a good guard over him, and the Government sets the price of wages, you may get something out of him. Otherwise, he would be an absolute failure." Richard Adams knew Mexicans who had been brought to California by a labor contractor to work for $1.10 a day and board. Another firm made Mexican immigrants work for ten and twelve hours a day, "handcuffing them at night to prevent their escape." Simon Lubin related similar violations of human rights when he quoted growers who said that they kept Mexican men "'at night behind bolted gates, within a stockade eight feet high'" so they could not leave.[121] That account of Indian labor in Los Angeles seventy years earlier resonates: "no longer profitable when cultivated with honestly compensated labor."

Cheap labor was at hand in the 1920s. The government allowed for the importation of Mexican immigrants. Later, whites from the southern Great Plains began to arrive as the farm depression became the Great Depression. Workers were plentiful, along with strikes and violence. Simon J. Lubin watched every move the growers made in these years and demanded that they justify their actions. After the Wheatland riot it was Lubin and his commission that proposed higher standards in the housing of migrants to reduce the possibility for violent confrontation. Fourteen years later, Lubin reported little improvement. When he confronted the convention of 1927 he did not urge clean working environments; instead, he attacked the system at its foundation: "If we have types of agriculture that can not be served by 'white men' as Mr. Frisselle calls them, by men who can take a part in the development of our culture, but [if instead we maintain types of agriculture] that demand amenable docile slaves, then we should do one of two things. We should so mould or remould that type of agriculture that civilized men can work in it, or we should send it to hell." [122] Lubin identified specialized agriculture itself as the root cause of the suffering he had documented in the orchards and fields of California. He recognized cheap labor as a structural support for a building that would cave in without it. Lubin still believed in good wages, sanitation, and "a scientific system of labor distribution," but he did not believe that California agriculture in its present form was capable of dignity.

California growers did not abandon the fantasy of a white man's California, they simply redefined it to mean the dominance of white growers over a labor system that used poor people to harvest specialized crops. Calls for a white labor force continued into the 1920s. "If this is to be kept a white man's country," said one writer, "white labor should be used in place of dark under every possible condition." A letter to the *Pacific Rural Press* in 1921 asked, "Isn't our agricultural edifice on a pretty precarious foundation when it is so largely dependent upon imported foreign labor? Can anything good be predicted as to its future?" The editor responded with a candor that would not have been expressed a decade earlier: "You are wrong in speaking about the 'disappearance of the indigenous' laborer. Such laborers in any quantity comparable with the demand for farm labor never existed in this state. . . . It is unfortunate that we have not a supply of farm labor such as they used to have in the older states, but we cannot claim to have lost what we really never had." [123]

In a terrible irony, a class of poor whites began to appear in the state. Rural poverty all across North America, a consequence of the unfolding crisis that began with the postwar crash in commodity prices, sent farmers adrift. Back in the 1880s observers of the labor market argued that if growers wanted to attract white labor they needed to provide decent accommodations and good money. Only impoverished immigrants would accept what little agriculture offered. But beginning in the late 1920s and increasing in the 1930s, many thousands came to California from the southern Great Plains looking for food and shelter anywhere they could find it, accepting the growers' work for the growers' wages, giving employers the greatest buyer's market they had ever seen. Cheap labor had no special color. [124]

6

Natural Advantages
in the National Interest

What natural advantages made possible, the institutions of industrial agriculture made manifest. California's rural industry turned on a principle: the careful matching of climate and crops provided the basis for an agriculture that regarded the world as its market. In Oliver Baker's conception, rails and wires gathered up the regions of North America to form one countryside. The products of various climates and soils met in the same markets. As a result of that competition, farmers examined and then exploited their local natural advantages, sorting out the arable lands of the continent according to their most profitable specialties. Just as each part in a machine has only one function, so too did the unity of North America as an agricultural economy demand that each region find a purpose to dedicate to the whole. The success of this project confirmed that an isolated region could trade in delicate crops even with many costly disadvantages. Difficulties that once forced farmers to search out the horizon or give up hope of ever reaching a city market had been reduced to debits in an account book. Those debits might still rise up and overwhelm the earnings of an orchard, but this was just the tension between advantage and disadvantage that the institutions of industrial agriculture were established to maintain.

Yet fruit growers had more than nature to thank for their fortunes. Almost every aspect of the business required some institutional counterpart to protect or enhance it. Indeed, industrial agriculture implied something more than crops and tillage keyed to an environment. It called for a tangled web of private and public authority and a latifundian division of labor. The Los Angeles Chamber of Commerce celebrated the plenty of California by praising the natural and "acquired" advantages that made it possible. Wrote George P. Clements, "Natural advantages are told in climate and soil—transportation and markets are of no less importance." Growers had acquired an integrated network of fast rail service, scientific implements, government regulation, labor procurement, and cooperative marketing—all to prop up a crop system that would collapse without them.

Full membership in American industry brought renown and strain to the fruit districts. When the federal government looked to the unirrigated lands of the Central Valley to be the beneficiary of a monumental water project, the business interests debated the idea. The farm economy lay in pieces after 1920, and growth for its own sake had lost its old glow. Yet it occurred to others that only continued expansion could save the fruit business from the competition its producers most feared. As they entered the national economy, growers entered the culture as well, pictured as Sunbelt millionaires and brutal reactionaries. California's national distinction, detailed in the present chapter, makes plain that although its regional agriculture had found a place in the wider world by the twenties and thirties, it had not found a point of rest. Any sign of stability was an illusion in a landscape never without conflict and flux.

The Contradictions of Growth

Impressed with California's contribution to the agricultural wealth of the United States, the federal government made known its interest in the state's continued expansion. Herbert Hoover, in particular, believed that the Far West's greatest achievements lay in the future. As secretary of commerce, Hoover told the Sacramento Chamber of Commerce in 1925 that the United States desired all the peaches and pears California could grow: "If we were to scan the whole nation for the greatest opportunity of national development we would find it lies right here in the Great Central Valley of California."[1] At the time of Hoover's speech 4 million acres out of a total of 14 million received irrigation, and some estimated that 6 million more could be brought under the ditch if water from elsewhere could be obtained. Six million acres, Hoover believed, would add "fully a million people to the state, and a wealth of one or two billion dollars—an addition to America as great as the fine state of Maryland." "This is no dry statistical statement," he added for accent. "There lie in its possibilities hundreds of thousands of the happiest homes in the world."

Hoover, however, was motivated by something more menacing than happy homes. As the former head of the wartime Food Administration he had a profound sense that today's fat might become tomorrow's famine. Hoover dedicated himself to the creation of plenty with hardly a moment's consideration that plenty had a different meaning to rural people in 1925 than it did in 1915. To understand his position and the motive for unceasing growth in the California countryside, we first have to know more about national agriculture in the 1920s.

By the time fruit growers had founded their corporations and secured their labor, these problems no longer concerned people like Oliver Baker. The fluid commerce between the sections, such an issue only twenty years earlier, became as ordinary as boxcars and telegraph poles along a country highway. In the meantime, the volume of food Americans enjoyed threatened the very farmers who created it, a fact that suggests the contradictions of agricultural growth. The farm depression that struck in 1920 presented a series of conflicting images. A remarkable change had taken place. The great crisis of the countryside—the one in which farmers failed the cities and choked off the ascending steam of industry— never came to pass. In 1929 Baker revised his story about a rising population and a declining food supply to reflect the achievements of the new agriculture: "Instead of population pressing on the food supply, as was feared a few years ago, the food supply is pressing on population in our nation, and appears likely to continue to do so."[2]

The proliferation of food between 1917 and 1926 had been the most astonishing in American history, all the more so because rural districts appeared to be losing population and falling into decline. In a brief nine-year period, 4 million people left the occupation of farming, 19 million acres went out of production, seventy-six thousand farms ceased to exist as farms, and agricultural production *increased* 25 percent.[3] The population of the United States grew by only 8 percent in the same nine years. In other words, agricultural production increased three times as fast as the population.[4] There were fewer farmers, but they were producing more food than ever before.

Abundance became an affliction to rural people. The crash in prices destroyed the farmer's buying power and threw agriculture into depression. During the period of high prices a $60 suit of clothes could be purchased with the product of a well-tilled acre—about forty bushels of wheat. After the decline it took two hundred bushels to buy the same suit.[5] Farmers struggled to make their crisis known to Congress, which at first ignored their claims. As Senator Arthur Capper of Kansas explained, "The general public and the majority of the Congress had not realized that the nation had passed into a new economic era in which the balance between agriculture and other industries must be more carefully safeguarded."[6] As farmers again tasted rancid hardship, urban Americans ate from a bountiful table, oblivious to the full implications of low prices. Hard on the heels of their short-lived affluence came bad times in which farmers suffered for doing exactly what city people expected them to do: make cheap food. Cheap food supported the American standard of living—cheap food helped to win the Great War, in the rhetoric of the time—but it left farmers scrambling for income after the peace.[7]

California managed to escape much of the damage as the house of agriculture caved in. The popular press wrote about the state's enduring prosperity—a stable front of affluence as predictable as a July afternoon in Fresno. "Even when most of the rest of the country was marked by the darkness of depression or the lighter shadows of uncertainty," remarked the *Literary Digest*, California enjoyed "amazing growth."[8] Which brings us back to the purpose of Herbert Hoover's visit to the Sacramento Valley. The secretary of commerce was in the state stumping for the Swing-Johnson Bill, legislation that would transform the Colorado River into a regulated waterworks, including a dam across Black Canyon that would one day bear his name.[9] But here is the point: in 1925, after wartime production had come back to ruin farm families from coast to coast, Hoover was talking about the need for more food. Did anyone want 6 million more acres of intensively cultivated ground to enter the agricultural economy? Hoover answered that the project he contemplated would require twenty-five years to complete. By that time, the population of the United States will have increased by 40 million— straining the food supply. "All this intensive production will be needed with this increase of population, for the easily cultivated land in the United States is today all occupied and our future supplies of food must come from just such projects as this."[10] Even when American farmers found themselves on the wrong side of prosperity, the political leadership still assumed that the problem of the countryside was scarcity.

Hoover argued that irrigation projects did all kinds of wonderful things. They concentrated the rural population, created more food, and brought a higher standard of living to farmers and consumers. These

had been some of the assumptions behind government-sponsored irrigation development since 1902, when Congress passed the Newlands
Act to establish a fund from the sale of the public domain that could be
used by farmers to reclaim desert lands. Promoters of big irrigation assumed that dams and canals increased the tax base and paid back investors
many times over the cost of construction. It was on these assumptions,
in addition to the faith that artificial water transformed useless desert into
blessed soil, that brought the state of California to propose (and the
United States to fund) what became the Central Valley Project.[11]

Growers backed government reclamation with Hoover's fervor.
Among them was Harriet Williams Russell Strong, the walnut and pampas grass grower from Whittier, who took up the reclamation issue during the world war. As chair of the Commerce Flood Control and Conservation Committee of the Whittier Chamber of Commerce and as a
member of the Arizona and California River Regulation Committee
(part of the Los Angeles Chamber of Commerce), Strong positioned
herself close to the center of western hydroelectric boosterism. In 1917
she came out with a plan to harness the Colorado River by damming the
Grand Canyon and claimed that the United States could pay for the war
by selling electricity. Years before, she had invented a system of dams
that reduced soil erosion in streams and irrigation canals, for which she
secured patents in 1887 and 1894. Strong wanted to control the Colorado River and its tributaries just like she tamed the streams behind her
ranch. She imagined the Arizona desert covered with walnut trees and
pampas grass thanks to the wise management of water.[12]

Strong knew, however, that the time had long since passed when lone
capitalists could finance water development. As she wrote in 1916, the
movement to reclaim the deserts in the name of the greatest good for
the greatest number had finally "reached the large canyons where the
chief and expensive work must be done."[13] Strong appealed to President Woodrow Wilson and canvassed Congress with all the high-minded
public spirit of a Roosevelt Republican: "It is the duty of the Government to handle the floods. No individually owned capital can do this."[14]
Only government could accomplish the miraculous alchemy of land and
water for "the welfare of all the people."[15] With the support and influence of growers like Strong, the industrial interests in California secured
the passage of the Swing-Johnson Bill in 1928 and had already written a
State Water Plan by 1931 to bring the rivers from the northern counties
to the southern deserts.[16]

But in what might have been their greatest hour, the years when expansion-minded progressives like Strong and Hoover imagined gigantic projects, all of their assumptions came under attack. Economists called irrigation development in a time of plenty a mistaken policy. An economist for the USDA labeled as "fallacies" arguments for reclamation that predicted higher incomes, increased taxable wealth, the necessity of government investment, and a rising population hungry for cheap food.[17] Most damaging to Hoover's ideas were statements like this one disputing the profitability of irrigation: "In recent years private capital has avoided irrigation development not because of the size of such undertakings but because of the absence of the prospect of reasonable financial returns."[18] Even the University of California Agricultural Experiment Station experienced a loss of faith, writing in 1930 that, although irrigation could be linked to the "entire commercial and industrial development of the state and nation," too much of a good thing could "impede the progress of the State in the future."[19] The university worried that more water would contribute to an agricultural depression that had, so far, left California better off than other regions. At a time when cooperative marketing associations invested in elaborate advertising campaigns to increase consumption and save their orchards, the last thing they wanted was more fruit to sell. Given the present status of agriculture, concluded the experiment station, "development which would involve bringing substantial areas of additional land under irrigation is not needed at this time."[20]

Oliver Baker, who joined the Bureau of Agricultural Economics in 1922, soundly thrashed Hoover's assessment of population and the food supply. The world had changed since the Great War. New tools now allowed fewer farmers to take more food from a smaller portion of the continent. Baker's list of important innovations included the transition of pasture lands into crop production after automobiles replaced horses, high-yield plants and animals, and an increase in yields per acre.[21] Farmers had succeeded in providing "more food and better food for the increasing population without any increase in crop acreage." Plans for expansion at such a juncture only confounded an economy on the mend.

Yet Baker saw the same foolish boosterism that sold off the Great Plains in the years before the war still operating all over the arid West. Amazingly, farmers continued to seek greener acres in the most unlikely places at a time when the United States was awash in marginal land. Even in the depths of the price fall, the worst depression of its kind in American history, with 20 to 30 million acres of land idle between 1920 and 1924, "over 4 million acres of forest and cutover land were cleared and made suitable for crops."[22] The hortatory speeches surrounding the clearing of fresh land—no matter its quality—put farmers under the mistaken impression that they would be able to make a living. In this light, a plan to transport the volume of entire rivers to millions of never-before-cultivated acres seemed destructive indeed. Stated Baker: "The artificial stimulation of agricultural settlement during the past half century . . . has been one of the fundamental causes of the present agricultural depression."[23]

The business boosters that Hoover met at Sacramento had their own reasons for supporting continued development in the 1920s. By that time, the rivers flowing from high in the Sierra to the floor of the Central Valley had been drawn upon to the point that they no longer furnished a source for expansion. Not that the cold water failed to run, but growers and boosters finally caught sight of its limit. Moreover, the need for continued reclamation seemed to increase as agriculture crossed the San Joaquin River and moved into the arid west valley. The technology that enabled this settlement was the electric water pump. Products like the "Pomona Turbine Pump," manufactured by the Lindsay Pump and Machinery Company, liberated growers from riparian sources by pulling water from deep deposits and spilling it onto the land.[24] Maps depicting isolated quarter sections divided into tracts removed from any river or canal indicate the use of groundwater in 1926.[25] By the end of the decade, growers on the west side irrigated four hundred thousand acres by pumping. When drought conditions in the years before 1924 made water deliveries all over the state uncertain, those with electric pumps fared better than those who still depended on mountain streams. C. C. Teague called himself among "the fortunate few" whose wells operated at normal capacity, even in dry years. Teague pulled five hundred thousand gallons per hour, or 12 million gallons daily, from the ground.[26]

Prolonged pumping lowered the water table in many places and a shortage of rainfall during the decade prevented its replenishment. "Year by year the ground water has receded, the cost of pumping has increased, and many orchards have been abandoned," reported the staff of the Agricultural Experiment Station, "if additional water is to be supplied to this area it must be conveyed from distant sources."[27] The California Development Association made the point that the problem could not be solved by a more extensive distribution of existing water sources. The irrigated districts had finally "passed the point where supplies of water from the watershed and drainage areas above them are sufficient for their immediate irrigation needs."[28] As the decade closed, groundwater depths in the southern San Joaquin fell to one hundred feet below the surface near Bakersfield and seventy five to one hundred feet in western Fresno County. High salt concentrations made the water in some places useless for irrigation.[29] When their pumps no longer raised a flood, residents of the west side cried to the state to help them maintain land already improved and in production.

Since the state and the growers refused to be hampered by temporary setbacks in the economy, great projects enjoyed wide approval and the confirmation of "necessity," regardless of how they might increase the amount of food in the economy and lower prices. Maintaining an increasing level of production had become more important than maintaining farmers in business; the development of the countryside had become more important than the countryside itself.

Shifting Scale

The increasing riskiness of agriculture—its costly inputs and narrow margins—forced changes in the scale and intensity of cultivation. Farms became larger, but for many reasons and not because inescapable economic forces commanded bigness. The large-scale farm had its admirers, some of whom seized upon bigness as the long-awaited fusion between agriculture and urban production that would bring prosperity to the countryside. California represented the top of this trend.

No single factor or cause created large-scale farms, but one point of origin might be the diffusion of practices and technology from experiment stations to farmers coast to coast. When farmers in different regions organized their countrysides with the same institutions that California growers used to organize theirs and realized the same benefits in marketing networks, insect control, and labor management, they threatened to beat the orchard capitalists at their own game. Each region may have had its own natural advantages, but each could use the same methods to till and trade. The marketing cooperative, as an example, became popular among wheat and tobacco farmers during the 1920s for some of the same reasons that pear growers adopted it years earlier. Growers on the Pacific Slope did not invent the cooperative, but when it came to the distribution of agricultural commodities, companies like Sunkist, Blue Anchor, and Sun-Maid redefined the possibilities of cash-crop farming.

California counted 255 organizations that marketed fruits and vegetables in 1925. By comparison, Florida had 98 and New York had 84 in the same year. California cooperatives claimed an estimated 42,322 members, while the three mid-Atlantic states claimed a total of 8,669 members.[30]

These numbers were bound to change with organizations like the American Institute of Cooperation. Founded in 1925, the Institute held its first meeting at the University of Pennsylvania where Henry E. Erdman of the University of California shared his thoughts about cooperative marketing with Secretary of Agriculture William M. Jardine, Edwin Nourse of the Institute of Economics in Washington, D.C., and Governor Gifford Pinchot of Pennsylvania. Newspapers and magazines also promoted marketing associations. *Wallaces' Farmer* carried stories about the California cooperatives, calling their combination of corporate organization and good old-fashioned working together the solution to the Farm Crisis.[31] One report to President Warren Harding in 1922 praised cooperative marketing as one of California's "most valuable contributions to the Nation" and claimed that "the cooperative movement, as developed in this State, has been spreading rapidly all over the country."[32] Wrote the *American Review of Reviews*, cooperation in "California has led the way," turning rural districts into healthy and wealthy suburban utopias with "pleasant homes, universal flower gardens, happy families, good society, delightful rural schools, good roads, with automobiles for everybody."[33] Comfort and affluence became the very expression of cooperation and suggested a way that farmers from Atlanta to Portland might finally redress the inequity between the countryside and the city. "The prevalent life in California country districts," wrote the editors, "has no apologies to make to the people who dwell in cities."[34]

But good public relations had another side. George Hecke, the director of the state Department of Agriculture, saw these developments as the divulging of California's formula for success. During a trip east in 1926, the director noted that methods of standardization and packing common in California had earned the admiration of producers in other states, and they seemed to be catching on. As he complained to Frank Honeywell, editor of the *Pacific Rural Press*, "Avantages which we have heretofore enjoyed are being decreased by the fact that other states are adopting the methods followed by California, and it will become necessary for us to make strenuous efforts to remain in the forefront in this valuable work."[35] Two years later, Hecke again confided in Honeywell his concern about the "increasing competition" that fruits from California encountered from other regions. "Such competition is unfair to our fruit industry on account of the agricultural disadvantages under which we work, such as higher paid labor, land values, permanent improvements."[36] Pacific fruit growers must keep pace with these upstarts, he said, or they would again feel the sting of the disadvantages they had worked so long to control. Staying ahead of the competition meant keeping the cost per unit low and productivity high; it meant new campaigns, new chemicals, new political power; it meant that California agriculture could not rest at any level of production.

Economists have a word for this kind of perpetual investment in agriculture—the "treadmill." This simple theory begins to explain the forces that made a few big farms out of many small ones all over the United States. Leading farmers take up a new practice or technology in the search for lower unit costs and higher yields. At first they enjoy generous returns, but it does not last. Writes the economist Willard Cochrane, "As more and more farmers adopt the improved technology, total market supply will increase. In a free market situation, other things remaining unchanged, this increase in supply will cause the price of the product to fall."[37] When only leading farmers innovate, the addition food they create is too little to have an affect on price. Then comes the rapid spread of the new machinery, organization, insecticide, or plant variety, and the market heaves. Consumers find more food for a lower price as operators watch their early gains disappear. Farmers who never bothered to buy into the technology, or who could not afford to, finally foreclose, unable to stay in business at the lower price. And after the innovation becomes common practice, leading farmers again go looking for something new to increase output or lower unit costs. This theory articulates George Hecke's concerns. When fruit growers in Michigan, for example, organized into cooperatives, acquired lead arsenate from the Ortho division of the Chevron Corporation, and hired Mexican workers for the harvest rather than take in hired men for the season, California growers felt the pinch of competition.

These pressures caused a shift in the intensity and scale of cultivation. To protect themselves against rising costs and falling prices, growers spent more and went big. Richard Adams, professor of farm management at the University of California, reported on the emergence of large-scale farming in the West.[38] Various forms of organization appeared in the 1920s, but the most successful employed salaried managers and service teams to keep machinery in top order, and organized its holdings into separate units. In 1924, 8,275 farming corporations filed income tax returns with the Internal Revenue Service; 9,309 filed in 1926. The firms raised grain, fruit, livestock, and cotton. Their total net worth (compiled from a set of 7,014 corporations) summed to $980 million in 1924.[39] "The average corporate farm," stated Adams, "has perhaps 20 times as many assets as that of the average farm, 23 times the net worth, and produces 26 times as much gross income."

The Bureau of the Census conducted its own study and determined that any farm with a product value of thirty thousand dollars or more in 1929 made the grade for "large-scale" designation. This value corresponded to a minimum size for a large-scale fruit orchard of from eighty to one hundred acres of bearing trees. The typical large-scale fruit farm was ten to fifteen times the size (meaning acres harvested, value of product, labor hired, etc.) of the average fruit farm.[40] California growers made an unmistakable contribution to the bigness vogue. In 1930 the Bureau of the Census evaluated 7,875 large-scale farms of all types in the United States, including 2,892 farms in California, of which 1,157 raised fruit. In other words, California's large-scale farms represented 37 percent of all such farms in the United States, and its orchards and vineyards alone represented 15 percent. The fruit business in California contributed more to the total number of big farms than any other type of agriculture in any other state or region of the country.[41]

In a sense, treadmill expansion offered nothing shockingly new in the 1920s. Just as fecund valleys once embodied the wealth of the "free land" frontier, bigger crops and lower costs provided the same elusive goal. For those with an ambition for surplus, intensive and extensive farming were not so different after all. Farmers chased the dollar in much the same way that they had once chased the horizon; only the tools had changed. What was new was that farmers could not set the speed of this expansion. The experiment stations came out with the latest indispensable aid to cultivation, which in turn demanded capital, which drove farmers closer to the edge of solvency. Growers who purchased inputs that they considered both essential and too expensive arrived at a crossroads. They could move to better land to realize higher yields; they could buy more land to bring in more money to finance the new technology; or they could sell out to neighbors caught in the same dilemma.

As Wendell Berry once observed, the technology-driven treadmill prods farmers to move at an unnatural pace, to cultivate according to the dictates of the economy and not according to the capacity of their land. Desperate for larger yields, the industrial farmer races to adopt whatever the experiment station invents, "having simply no choice but to continue to do as he had done until the bank closes him out. He cannot change because he has no margins to turn around in. At this point he is literally dancing to a tune called by an economy that has always (as he sees too late) proposed his failure as the price of his participation."[42] Those who failed to keep up fell off the speeding belt. In this way, large industrial farms tended to eliminate small-scale farmers who could not find stability at ever-lower prices and ever-higher costs. The treadmill changes the face as well as the shape of the countryside.

Yet California's shift in scale is more complicated than its apparent rise to the top of this trend. By 1929 the state counted 5,735 fruit farms with between one hundred and one thousand acres or more, but it also counted 46,274 farms with less than ten and up to ninety-nine acres. More than half of the total number of orchards and vineyards had between ten and forty-nine acres.[43] These numbers suggest an unusual circumstance in agriculture: a mixed regime consisting of very small and very large properties. Why did the irrigated districts support groves of vastly different sizes? The answer brings us back to the nature of intensive cultivation.

Growers had all kinds of reasons to keep their plantations small. The absence of machines for harvesting meant that the fruit business enjoyed none of the economies of scale that derive from mechanized production. "Economies" in the orchard came from the concentration of materials and labor on small pieces of land and the expected high yields that would distribute high costs over an ever-greater number of units. But although growers first perfected intensive cultivation on tracts of twenty and forty acres, no law of economics prevented them from employing the same methods on two hundred or four hundred acres. Simply put, just because it could claim no machine-driven economies of scale, it did not follow that fruit growing harbored *dis-economies* of scale. Orchards could take on giant dimensions as long as they expanded with one key ingredient: management.

Managers did more than stand in for owners while the latter dabbled in San Francisco politics or vacationed at San Diego. The manager worked to extend the same degree of investment and attention that the owner achieved on the home plot to more or larger units. Management sought to duplicate the intensive methods of small-acreage cultivation many times over, either by operating several farms in different places or by dividing great tracts into manageable units. In this way, the number of acres that could be tilled under one owner had no limit. The economist Edwin Nourse recognized the manager-engineer as the first agricultural professional. Nourse said that the manager's function should be strictly differentiated from the common "operative processes of the farm" and that only by entrusting key decisions to their input experts would farmers "take advantage of the technological possibilities which lie in the path of modern scientific and commercial agriculture."[44] Though Nourse admitted that "farming is not a factory occupation," nonetheless he admonished that "we shall make better progress by attempting to see *just how far* the scientific management point of view will go in agriculture."[45]

By 1930 it had gone a great distance in California. Managers operated 7,768 farms in the state, or 6 percent of the total, up from 4,949 in 1920. And managers tended 5,869,028 acres, or 19 percent of the land under cultivation in the state, up from 5,485,447 in 1920. Just three counties included 33 percent of the managed farms: Fresno (939), Imperial (613), and Los Angeles (985). Fruit growers, in particular, dominated this trend statewide. Of the 52,009 fruit farms, 4,141 operated under the care of managers. That means that less than 10 percent of the state's fruit farms represented over 50 percent of its managed farms.[46]

The rise of management in California and elsewhere came along with warnings and arguments from experts against small-scale farming. As the number, complexity, and expense of inputs increased, small-acreage ventures ran into cash-flow trouble. The Agricultural Experiment Station believed that although many farmers produced for a cost well below the selling price, those with modest holdings often failed to generate enough income to pay for operating expenses, interest on loans, and the high standard of living many had come to expect. Frank Adams of the California Extension Service called agricultural businesses "too small in many instances to meet the demands made upon them when prices drop or expenses continue at high levels." He predicted a "readjustment in the sizes of acreages," in which some owners would sell their holdings to better located or more skillful neighbors.[47] Even a peach orchard yielding a healthy ten tons per acre with sales at twenty-five dollars a ton might not turn enough cash to keep the business going if it counted only twenty acres. "After all," concluded Wheeler McMillen, associate editor of the *Country Home*, "the objective we seek is not net income per acre, but net income per individual farmer."[48] A small orchard might maintain its owner in business but might not maintain the pool of available capital that would keep its owner up and running on the agricultural treadmill.[49] Largeness might be less profitable per acre, but it was more lucrative overall.

Yet questions about large-scale agriculture lingered. Richard Adams, who wrote about the corporation farm in the West, never proclaimed its advantages. In fact, Adams saw few advantages in big farms, called them risky, and cautioned that they depended on good management. Giant operations simply appealed to a certain entrepreneur, he explained; they were no more profitable per unit than small farms.[50] J. Karl Lee, an economist writing in the 1940s, compared medium and large farms in the southern San Joaquin Valley. He found "no clear-cut margin of efficiency among the various size farms." Lee added, "There are obviously advantages to all size farms, or some would have been eliminated in favor of those sizes which have a definite advantage. . . . If efficiency is interpreted in the usual narrow sense—output of money per unit of input— then the large farms are the most efficient. But if efficiency is regarded as the return to all parties concerned and profit per unit of land, then the medium-large and medium-size farms are the most efficient."[51] Even Edwin Nourse conceded that "the degree of efficiency attained on the big power farm under something resembling factory methods can be practically duplicated in a very large part of our well-developed farming country at a cost as low or lower."[52] But never mind these subtleties: the factory farm had arrived, and to a coterie of business people, politicians, and economists it embodied all that was progressive, modern, and affluent.

Large-scale industrial agriculture appealed far beyond the Pacific Slope because it suggested a manner of farm organization fit to survive the violence of the national economy—a solution, some said, to the question of agricultural prosperity in the 1920s. If bigger meant more efficient, and if efficiency offered a kind of "weatherproofing" against the storms always off the coast of capitalism, then agriculture may have finally found a way to stability. Advocates of the modern farm denounced the old family farm as deficient, archaic, and incompatible with an economy based on high technology and world trade. They stressed the scientific control of nature and the superiority of economics over the unschooled ways of common tillers. The policy they favored insisted that agriculture function as a sector of the manufacturing economy. Most of all, the "modern" farm emerged as a creature of the nation's demographic transition from rural to urban in the 1920s and of the depression of the same decade that cast poverty across the American countryside.

Expressed in terms of the politics of the day, big-business agriculture did not need farm relief. An important context for the advent of the modern farm as the solution to the rural problem is the debate over the McNary-Haugen Bill of 1924—the first legislative plan to help farmers after the great price decline. The objective here is not to argue something about the bill itself but to explain how a set of industrialists used their opposition to farm relief as a springboard to propose alternative ideas about the future of American agriculture. The McNary-Haugen Bill proposed a two-price system for crops, in which a corporation founded under legislative charter would purchase farm commodities for sale to the public at a high or "fair value" domestic price, while any surplus would be diverted for export at a lower world price.[53] Supporters argued that the health and security of farmers should be part of the price of food. When city people paid more for their bread, they helped to secure economic parity for rural folks.

President Calvin Coolidge bristled at the idea that government should help farmers. He struck the bill down in 1927 and again the next year. In a speech before the American Farm Bureau Federation in 1925, Coolidge said that the American farm needed all kinds of inputs but no one's help: "It has become a great industrial enterprise, requiring a broad knowledge in its management, a technical skill in its labor, intricate machinery in its processes, and trained merchandising in its marketings. Agriculture in America has become raised to the rank of a profession. It does not draw any artificial support from industry or from the government. It rests squarely on a foundation of its own. It is independent."[54] Coolidge assumed that farmers required little more than land on the breaking and the know-how to make it productive of capital. Though manufacturers sold their products under the protection of a tariff—itself a two-price system, one for domestic consumers and one for dumping surplus abroad—agriculture had to make it alone. Farmers purchased inputs just like factory owners and needed financial institutions, transportation companies, labor contractors, and advertising to sell their commodities. But as long as they did not need government they remained "independent."

Coolidge also sensed the changing composition of the public when he turned against farm relief. Finally almost equal parts of the population, city and countryside, sat on different sides of the industrial economy, and Coolidge knew which side kept him and his party in office. Time and again in his long veto address he pointed out the consequences of farm relief for the people who did not farm. "It is not inconceivable," he threatened, "that the consumers would rebel at an arbitrarily high price and deliberately reduce their consumption of that particular product." Because the bill amounted to "a direct tax on certain of the vital necessaries of life[,] it represents the most vicious form of taxation."[55] Most Americans did not need to be convinced. The McNary-Haugen Bill failed utterly with the public outside of the Middle West, and Coolidge's veto brought an enormous response from the East and Far West. Never before had a single issue provoked as many letters and telegrams to the president as McNary-Haugen—a deluge, said the *New York Times*, of "commendatory messages." A number of the telegrams praised Coolidge for holding down the cost of living—for acting "to protect the consumer" rather than giving farmers a guaranteed income. Supporters called the veto reason enough to back Coolidge in 1928.[56]

The industrial farm was not the only alternative to farm relief. As we have seen, cooperative marketing in California and elsewhere made fast gains after World War I. The cooperative, however, did not always make good its promise to give farmers a modicum of control over prices. Voluntary restrictions on production to bring prices up to parity asked farmers to grow less, sell less, and still stay in business through the worst price decline in American history. The demise of McNary-Haugen appeared to mark the demise of a political solution to the crisis, and in the vacuum that followed the veto some opponents of farm relief suggested a corporate model for agriculture. These writers understood that technological and organizational change implied corresponding social change—that modern farms meant different and fewer farmers. Though they did not write with California in mind, their calls for a rural industrial elite as the rootstock for a new countryside evoked the fruit growers.

Conservative thinkers assailed farm relief. Robert Brookings, former president of Washington University in St. Louis and founder of the Brookings Institution, refused to acknowledge the differences between farming and manufacturing and claimed (without offering good reasons) that any successful farmer could overcome these supposed differences and rule his land as Henry Ford ruled his assembly line. "The one great source of 'farm relief' consists, not in raising prices, but in encouraging those sound business methods in agriculture which have developed many of our other industries into their world-dominant position of efficiency, such, for example, as the automobile industry."[57] Henry Ford himself had automobiles in mind when he wrote about the problems of agriculture. Ford called farming a "little business in a world of big business" and exhorted farmers to stop seeking government relief for problems that, he believed, were their *own* problems and not the manifestation of a contrary economy.[58] As Ford explained, "A farmer does not plough with money; he does not sow with money; he does not cultivate with money; he does not harvest with money. His problems are of production, not of finance."[59] To prove his point, Ford built a model of modern food production (with lots and lots of money) on several thousand acres at Dearborn, Michigan, where employees operated a dairy farm "exactly as though it were a factory." Machines fed, washed, and milked the cattle. At plowing time managers directed fifty tractors to move in unison across the vast property. Workers came from Ford's automobile factories and received the same wages paid on the assembly line.[60] Ford never explains how family farmers might implement his methods on their own land.

But perhaps this was never Ford's intention. As the farm depression persisted, economists and leaders in agriculture expressed a vision of the countryside that called for food producers to live or die by the same rules that governed profit and loss in the jungle of capitalist enterprise. Those rules had never worked to the advantage of farmers. Estimating demand, adjusting production, setting prices—the most basic functions of factory management—could not be replicated in agriculture. Yet the literature of factory farming from the late twenties and early thirties is ignorant of its own assumptions, haughty of its certainty, and dismissive of alternatives. The farm problem could be solved by the assertion of industrial power over a process that had for too long lay in the hands of people possessed of rusted tools and rusted thoughts, they said. The industrialists made no room for farmers on hard times, calling them small-minded and wishing them away into the city with an attitude that would have shocked reformers of the country-life era.

Edward Mead, a professor of finance, and Bernhard Ostrolenk, a former director of the National Farm School in Pennsylvania, expressed this general view when they wrote about the transformation of the countryside and its collateral damage. "It is impossible, and it has never happened, that an industrial revolution such as is now in progress in American farming, can be stopped or stayed by any scheme of public subsidy to the victims of the revolution."[61] The end result of this "unstoppable" movement would be a shift of all the unproductive people from the countryside, "where most of them are living on a scale comparable to that of the European peasant," to the cities. The authors called such a migration inevitable: "Nothing short of nationalistic control of farm operation in the interests of an ill-equipped, untrained, unintelligent, backward farm population will suffice to retain this population on the farm." Let there be no talk of increased taxes on the rich and no program to buy up farm surplus to maintain prices. Tinkering with the tariff also posed a threat to the many who did not farm. In fact, the authors eliminated any solution that would have called upon city people to give a little in order to mitigate the extremes: "Taken together, the demands of the rural population represent a serious danger to the interests of other classes." The United States was an urban country: "Its primary interest in the prosperity of the farmer is in the bearing of that prosperity upon the city cost of living. . . . It is, therefore, difficult to believe that the city population will allow an agricultural minority to increase the city's cost of living."[62]

Who would grow all the food America would need as it approached midcentury? Ostrolenk believed that the very crisis so dreaded by the advocates of farm relief would result in the emergence of a professional agro-industrial class—a set of superefficient farmers who could survive any panic, any weather, any depression and still turn out cheap food: "The same revolution that is bringing economic disaster to millions of farmers has brought prosperity to a competent group of business farmers, not exceeding 800,000, and constituting less than one-eighth of the whole, who are finding the new order profitable. These are the farmers who are adopting the new methods of production."[63] This industrialized eight hundred thousand, the Darwinian survivors of "treadmill" competition and the shifting scale of agriculture from coast to coast, would inherit the arable lands of the continent.

The new farm professional would be an agrobiologist, according to Oswin W. Willcox. A onetime professor of soils at Iowa State College, Willcox argued that the role of the modern farmer was not to till the soil but rather to "superintend the unfolding of the life of vegetable organisms." Willcox advocated a superintensive, highly technological agriculture fully aware that it would result in fewer farmers and perhaps "far-reaching economic and social struggle and readjustment." An economic line separated those who could and those who could not produce at a profit in the agrobiotic age: "It would perhaps be more appropriate to call this line a zone of social-economic wreckage where men pass out of economic existence."[64] The "new social order"—the "era of abundance"—carried a price. Perhaps it is no surprise then that when Willcox chose a political system to bring about his agrobiological utopia it resembled Italian fascism.

Willcox chided those lovers of liberty for whom individual rights always trump economic imperative, people who "fall into a rage over any proposal to impose even a diluted fascism on free American citizens."[65] If incompetent farmers refused to give up and sell out, then they would be removed by the state. So said an Italian law declaring, in the author's words, that "private property in land is a public trust to be held under an obligation to make a due and proper social-economic use thereof."

The Italian farmer learned that if he could not make his fields profitable he "must step aside; the general welfare will be served, and will at least indirectly subsume the welfare of the displaced farmer himself."[66] Of course this is an extreme view, and it might be easy to dismiss were not Italian, German, and Russian modernizers in power in 1934, the year Willcox published *Reshaping Agriculture*. Like them, Willcox believed that no nation could move into the "era of abundance" with a countryside that did not serve the ambitions of the state.

A strange dichotomy emerges from this (admittedly incomplete) discussion. Agriculture would live in total isolation from government or it would be subject to its total control. It would be autonomous or it would be planned. Neither view acknowledged the crucial position of farmers as the keepers of a sensitive ecological relationship—the very relationship that made all factory production and human reproduction possible. Neither view recognized the strange position of farmers in an industrial economy. The common factor unifying Coolidge and Willcox was the belief that farmers performed a critical function but composed no constituency. As the source of food, raw materials, and commodities, agriculture made industrialism possible; as the basis of a rural society, a regional politics, and a middle landscape, it had no importance at all.

The revolution in farming that began in California favored larger units, a more narrow variety of plants, expensive inputs, and a depopulated countryside. It did not spread as quickly as its supporters would have wanted, nor did it proceed without opposition, but it resulted in agribusiness as we know it—a term for the vertically integrated process of making food with as much technology and as few people as possible. As for farm relief, when it finally passed into law it ended up serving the same big business cultivators that Brookings and Ostrolenk claimed could live without it.[67]

Orchard Culture in American Culture

As it entered the economy, California agriculture also entered the culture. Who were these farmers from the state where movies came from, and how did they succeed? Fruit growers affected an attitude of big-business confidence and callousness. They were depicted as greedy little men and armed guards, and they were even personified as the state of California itself.

Since fruit labels first arrived at New York's Terminal Market, Californians have propagated an image of gracious landscapes and an uncomplicated affluence set against great mountains and gentle waters. Fruit growing itself—and especially the business of oranges—conveyed the brashness of a state whose farmers had escaped the difficulties that befell farmers in states like Kansas, Connecticut, and Mississippi. Writing in the *Saturday Evening Post*, Will Rogers depicted California as an upstart—the symbol for a new Sunbelt culture. Rogers staged a mock debate between Florida and California to determine the greatest state in the Union; the conversation quickly turned to fruit. Florida says: "We are known for our Oranges." To which California responds:

Why oranges, climate, Pickford, California—those four wonderful words can never be dissociated. I will admit there is a bootleg variety of orange that thrives up to the size of a green plum on the banks of your swamps; but as for being called an orange, that is only done, of course, through a sense of humor. We take Florida oranges to California, dry them and use them for golf balls. As for taste, they resemble the green persimmon.

> *Florida:* Our grapefruit sells for about $10,000,000 a year, and we think it is the best in the world.
>
> *California:* We use the juice of your grapefruit as a fly spray. We had no idea anyone ate them.
>
> *Florida:* Our oranges alone in 1924 brought us in $15,000,000.
>
> *California:* That would just about pay for the labels on the ones we shipped.[68]

Asked why it does not grow corn or tobacco, California answers that "[a]s for tobacco, we can't raise as good as North Carolina, and because we can't we don't raise any. No inferior anything for us. It's the best or none at all." Rogers's spoof contained a quiet truth: the two states really seemed to be vying for some title, trying to extend their influence to neighboring states, capturing fresh markets for their commodities. California had already sent its managerial, irrigated agriculture to colonize the Imperial Valley and across the Colorado River into Arizona. And yet the debate staged before the "Prevaricators' Club of America" alerted readers that what they heard from the Sunbelt might not be the whole truth.

Neither could American audiences trust Hollywood to accurately portray life in rural California. Produced on Los Angeles soundstages and in the dusty hills behind Burbank that doubled as the Wild West, the movies projected a stylized life and landscape. Below those hills, stretching out into the San Fernando Valley, lay hundreds of miles of orange trees that occasionally appear in the background as Charlie Chaplin or Harold Lloyd dash over dirt roads before spilling into Beverly Hills. In 1923 Hal Roach produced *Oranges and Lemons*, a silent film starring Stan Laurel (before his team-up with Oliver Hardy) as a hapless picker and packer. He wears a mock sombrero, he fuddles with equipment, he runs from the mean manager, he flirts with the girl, he makes a big mess.[69] But the film uses fruit growing as a mere premise for comic mayhem.

The high culture of the orange groves received its most notable depiction in 1934 when W. C. Fields made *It's a Gift*.[70] Fields plays Harold Bissonette, an ill-mannered and incompetent dry-goods merchant from New Jersey. When a wealthy uncle dies and leaves Harold a piece of property in Southern California, the Bissonettes sell the store and move. The family finds a worthless tract of desert land, its trees dead from drought and neglect, and its bungalow home wind-blown and broken. Just as his bitter wife threatens to leave with their son, Bissonette gets news from a neighbor that a developer wants to build a racetrack on his sorry grove. Bissonette holds out for forty-four thousand dollars and buys a thriving ranch with a substantial home.

In the last scene, the grower is relaxing in tennis whites under one of his trees. He plucks an orange and squeezes the juice into his whisky. His wife and son step into a waiting limousine. Nearby, a stack of boxes marked "Bissonette's Blue Bird Oranges" stands picked, packed, and ready to be shipped. The tennis whites and the whisky speak for the assimilation of old ways with new. Americans learned that Bissonette had not undergone a transformation when he came to California, nor had he bothered to learn a thing about the cultivation of oranges. He was the same buffoon that once suffered a comfortless existence in the humid East; he had simply lucked into the easy opulence of an irrigated life.

The buffoon that Fields portrayed failed to charm everyone. By 1934 Franklin Roosevelt had been in office for a year. The Great Depression and the disaster on the Great Plains brought people whose parents had once gathered buffalo chips in eastern Colorado and western Oklahoma to California in search of work and relief. In their eyes, the grower no longer represented the possibilities of a luxuriant and fertile land but the hardened heart of business and the paranoid politics of red baiting and union busting.

Simon Lubin, the progressive reformer, publicized the criminal activities of growers at a time when no one else did. Frustrated with the laggard pace of labor reform after so many years, Lubin attacked the rulers of the industrial countryside and informed the public of their tactics.

He spoke out in 1934 against the growers' most vicious attempt to shackle the liberty of their workers. In that year, S. Parker Frisselle became the first president of the Associated Farmers, an organization founded in 1934 after a battery of violent strikes hit the major cotton and fruit districts. The Associated Farmers took up the work of the Farmers' Protective League, that brief alliance of influential growers founded in 1914 in response to labor and wage reform. Its members dedicated the Associated Farmers to breaking the rural labor movement and any attempt at unionization. Growers in 1934 chose not to act through public suasion and the ballot box but through illegal arrests, stockades to hold strikers without charge, and vigilante night rides often conducted with the cooperation or the neglect of local police.[71]

Lubin recognized these tactics as staples of political terror and brought his anger to its source. He directed the following words to the members of the Commonwealth Club of California, the state's business elite:

They forbid free speech and free assembly—even by ordinance! Brutally they break up peaceful meetings conducted in private halls. They interfere with the organization of labor, and deport representative committees fairly elected. Indiscriminately they arrest innocent men and women under fake charges and, through the use of the suspended sentence, hold them in constant fear. . . . They browbeat and intimidate, and subject to physical injury, the attorney for the workers. They scoff at our Federal Courts. . . . Their so-called peace-officers do the bidding of their masters with the able assistance of pistols, machine-guns, tear-gas, bombs, hard-wood sticks—and a sufficiently large deputized rabble to supply the chorus for fifty comic operas. . . . Then they blame the consequent disturbances upon 'Communism' and 'Moscow!' Though the cause is immediately under their very eyes—a cause which in great part they themselves create—nevertheless they seek it literally ten thousand miles away![72]

The fruit grower had become a pitiless figure. The same people who came to colony tracts to build a suburban life had, within a generation, become the merciless tyrants of an industrial countryside. The survival of a dubious form of agriculture brought them to justify the worst violations of law and human rights.

The Associated Farmers came to national attention when a Senate subcommittee chaired by Robert M. La Follette Jr. of Wisconsin traveled to San Francisco and Los Angeles in December of 1939 to hold a series of hearings. His investigation into violations of free speech and the rights of labor set out to demolish the secrecy surrounding the Associated Farmers. First on the agenda: to subpoena the leadership and determine the extent of their activities. It proved difficult for the committee to find out just who belonged to the organization. The Associated Farmers operated as a federation of local associations with a central leadership that claimed little authority, almost in the manner of a cooperative marketing association. The structure seemed designed to protect important members from public scrutiny.

In one instance, Senator La Follette questioned Joseph DiGiorgio, president of the Earl Fruit Company and the DiGiorgio Fruit Corporation, about his financial support of the organization:

Senator La Follette: Mr. DiGiorgio, did you have a position or are you associated with any group or committee of the Associated Farmers of California?

Mr. DiGiorgio: No. As I said before, Senator, I never attended but one meeting since the Associated Farmers was founded in Kern County.

La Follette: I was not speaking of Kern County.

DiGiorgio: They have a local. You know, each county has a unit.

La Follette: I am speaking about the Associated Farmers of the State.

DiGiorgio: I am not a member or [in] any position in the Associated Farmers of California; but the local they have me [as their] financial agent.

DiGiorgio explained that he served in this capacity "against my wishes." But then revealed that he had raised ten thousand dollars in 1938 for the state organization to hire a public relations "reporter" and had communicated about the matter with Holmes Bishop, the president of the Associated Farmers. The committee learned far more about the federation from the outsiders who testified, including the economist Paul Taylor and the attorney Carey McWilliams.[73]

Inspired by Lubin's legacy of advocacy and confrontation, a group of young intellectuals responded to the Associated Farmers by establishing the Simon J. Lubin Society and the *Rural Observer* to fix a public eye on agriculture. They helped to reveal the Associated Farmers and their proxies, including C. C. Teague, George Hecke, and George P. Clements, head of the agricultural division at the Los Angeles Chamber of Commerce.[74] The editors of the *Observer* knew that farming in California belonged to big companies. So they followed the money to Calpak, Libby McNeill Libby, the Southern Pacific Railroad, Miller and Lux, Transamerica Corporation, the California Chamber of Commerce, and the Kern County Land Company. They gathered facts to establish that the land belonged to these same corporations and not to common people. According to the *Observer*, 2 percent of all farmers owned 25 percent of the acreage, paid 36 percent of the wages for hired labor, and received 32 percent of the total crop value.[75] There was nothing new about corporations operating in the California countryside, nor was the involvement of railroads and the chamber of commerce in agriculture a development of the 1930s. But it was shocking to liberals like John Steinbeck, Carey McWilliams, and Helen Hosmer, who edited the *Observer*. Implicit in their writings is the idea that the American farmer was the last independent worker, the final holdout against industrialism, and that big business farming aimed to take his land and turn him into a wage slave. The California countryside lay firmly in the grasp of the "corporate interests," said the *Observer*, and justice could only be achieved with a labor movement and class struggle.[76]

The struggle found a compelling voice with the publication of *Factories in the Field* by Carey McWilliams (1939). The book, a journalistic account of migrant labor, shattered the bright-sunshiny public image of rural California by repudiating its history and divulging its roguish activities. The book's accomplishment lies in its comprehensive argument and research and in its searing ethical voice. McWilliams called agriculture a front for political reactionaries and a snare for working people. The narrative reaches a climax with the strikes of the 1930s and the formation of the Associated Farmers. In "Gunkist Oranges" McWilliams relates various attempts to stop Mexican orange pickers from striking, including armed high-school kids and football players from the University of Southern California "masquerading as amateur storm troopers." "It is no exaggeration to describe this state of affairs as Fascism in practice," he wrote.[77] McWilliams also served the public as a defender of farm labor. Governor Culbert Olson (a New Deal Democrat) appointed the writer (then only thirty-three years old) to chair the state Commission on Immigration and Housing in 1938, the same position that Simon Lubin held under Hiram Johnson.[78]

John Steinbeck further defined this interpretation in the *Grapes of Wrath*. The brown, scrubby hills behind Needles, the first town across the Colorado River, told Tom Joad about the kind of life he and his family would find in the Central Valley. Here, on muddy banks, Tom first realizes that California does not mean to deliver what it promised: "Never seen such tough mountains. This here's a murder country. This here's the bones of a country. Wonder if we'll ever get in a place where folks can live 'thout fightin' hard scrabble an' rocks."[79] Pa says, "Wait till we get to California." Tom says, "Jesus Christ Pa! This here *is* California."

Immediately after crossing the river, the Joads learn that certain people rule the place. They meet a man and his son from the Panhandle who are trying to make their way back to Oklahoma. A tense conversation follows in which the two tell Tom and Pa about what lies over the desert—Bakersfield, back-busting work, and violence. The man tells Tom about the grapevines and the orchards, about the land that no one cultivates as it lies idle, about the people who hate common folks like them. He tells the Joads about armed guards who shoot any one hungry enough to take fruit for themselves and about the miserable life along the highways: "You ain't gonna get no steady work. Gonna scrabble for your dinner ever' day."

The faceless men with clubs and guns stood for a countryside all out of order. In Steinbeck's treatment there was no California for the Joads, no fortunate place where fruit hung heavy on bending branches, where natural advantages and the experiment station secured an independence. As the man from the Panhandle said of it: "She's a nice country. But she was stole a long time ago."[80] The country was stolen by the owners of all this misery, by people Steinbeck calls "little shopkeepers of crops, little manufacturers who must sell before they can make." Americans knew that corporations controlled city industry, but they still expected to find democracy in the heartland. As they crossed the river of disillusion with the Joad family, Americans learned that the farmers' republic no longer survived in rural California. Out of place and almost unrecognizable as farmers, the Joads become the hopeless proletariat of the industrial countryside, the uncompensated victims of the agricultural revolution.[81]

Progress had ceased to be progress. Intensive agriculture, which once offered a way to bring forth abundance from scarce farmland, appeared increasingly destructive of human rights and the natural environment. Growers gained in political power along with the importance of their industry but retreated from the centers of population. They no longer came to colony tracts as retired railroad engineers and schoolteachers but graduated from their own professional schools. As suburban Los Angeles and the San Francisco Bay Area expanded, agriculture became less visible to most Californians and more suspicious to them as well. The countryside landed on the front page or in the Sunday magazine as the location of unnegotiated strikes, homeless migrants, and chemical pollution. Farming had indeed been reorganized on the basis of "modern industrial efficiency," but it had not found stability.[82] Every solution posed a problem; every input upped the cost. Even as growers realized the fruits of natural advantage on millions of temperate acres, the elements of a flourishing agriculture remained beyond their reach.

Restless Orchard

By the 1930s the California fruit business represented industrial farming at its apex: The almost complete separation between farm production and consumption and the dedication of soils in a vast region to consumers far away. Though nature presented a set of ecological options making possible a great diversity, the growers' particular reading of nature led them to plant a limited number of plants in monocultural stands. Determined to enjoin California with the emerging national economy, they invested in labor practices, chemical inputs, and marketing organizations intended to sustain specialized crops. People and nature served the growers in a singular capacity, but the growers refused to serve either in return.

This has been a story of the middle landscape in California. Yet by the 1930s the boundaries of that landscape were not so easy to identify. Writing in the next decade, anthropologist Walter Goldschmidt went in search of the lines connecting families and communities in places where industrial agriculture reigned. What he discovered brought him to argue that old distinctions between urban and rural had ceased to hold. Goldschmidt studied Arvin, a town in the San Joaquin Valley, and there observed its transient population, its unresponsive newspaper, and its fractured community made up of laborers and white-collar workers. Urban need not apply to cities alone, he concluded, because impersonal relations, heterogeneity, and economic interdependence defined life in rural California. In short, Goldschmidt said of Arvin and the other towns he examined that their "degree of urbanization varies with the degree to which farm operations have become industrialized."[1] Like the country-life reformers of thirty years before, he favored rural institutions but discovered that California had few of them.

Goldschmidt's inquiry sends us back to the first pages of this book and the debate about the future of the American countryside that topped the political agenda for a brief time at the beginning of the century. There we met two thinkers who illustrated opposing conceptions of rural reform. By the middle of the century, Liberty Hyde Bailey and Edwin Nourse had conducted careers that reflected the career of industrial agriculture in America.

Liberty Hyde Bailey never endorsed the assumptions or the methods of industrial agriculture. When asked in 1927 to define "the measure of rural progress" he gagged on the prompt: "I do not like the question because I do not know what you mean to imply by 'progress,'" he said. So he recast it: "Let me assume . . . that you ask, What is the measure of rural welfare? Then I reply that there is one measure. . . . *It is the degree of satisfaction the farmer and his family derive from the occupation.*"[2] Yet the values that Bailey championed no longer inspired business, research, or public policy. The American countryside, far from representing a social and environmental alternative to urban life, became a place dedicated to growth for its own sake and an urban standard of living, a place sustained from without rather than from within. Bailey continued to insist that agriculture manifested the biological interdependence between people and nature and amounted to more than simply another sector of the greater economy. His moment of influence now past, Bailey returned to flowering plants, to the cultivation of evergreens, sweet williams, greenhouse petunias, and other plants in the ornamental landscape.

Yet although he turned away from the agricultural megamachine, he never lost track of it. In the preface to *The Garden Lover* (1928), Bailey observed that horticulture "is now largely visioned as a commercial undertaking only; its methods, and particularly the varieties of plants, must be standardized; products must be raised in quantity; the output must be reduced to rigid grades." He recognized the importance of horticulture's "commercial career" and appreciated the benefits of uniformity in the production of fruit. He reserved his praise, however, for the amateur gardener—"the ultimate conservator of horticulture." Bailey no longer asked the farmer to broker an agreement between human artifice and nature. That responsibility fell to the gardener for whom "interest lies in the individuality of plants rather than in sameness of specimens."[3] He considered amateur cultivation a declaration of independence from the bland process that furnished city people with their food. "The trend of events," he wrote in 1928, "is to discourage the amateur (the lover) and to cause him to give up his personal product for the greater output, precision and profit of the machine and the managed industry."[4] The gardener stood "against commercial valuations of life," sought a genuine experience with nature, unified production and consumption. Those lacking this vital contact with the earth bought their food in cans and became "standardized by the mere force of circumstances and imitation." Though his words called for no direct political action, Bailey certainly intended to undermine industrialism by encouraging people to take control of their food.

Edwin Nourse continued to advocate industrialism in agriculture. By the time of his death in 1975 he had helped to transform the countryside into an adjunct of the nation's export trade, just as he set out to do early in the century. When Nourse looked at rural lands in 1930 he no longer saw eighteenth-century modes of operation. "Scientific farming" had become common practice. Farmers integrated plant breeding, soil feeding, and insect control "into their everyday farming methods with a speed and effectiveness which are truly remarkable."[5]

Nourse's ideas did not echo through empty lecture halls, nor did he labor in obscurity to formulate them. He built a career as the agricultural economist who made no excuses for agriculture, and his influence seemed to increase as the rural depression worsened. When politicians wanted an analysis that placed the burden of change on farmers and not on Congress or the president or the corporations, they reached out to Nourse. He published *American Agriculture and the European Market* in 1924 to probe the international causes and implications of agricultural economics. Nourse never let up against the "old sentimental and habitual patterns" that he believed continued to infect farming with backward thinking.[6] He joined the Institute for Economics in Washington, D.C., soon to be called the Brookings Institution. President Harry Truman appointed him to chair the Council of Economic Advisors in 1946. Reporters gathered in Washington hallways to hear his opinions. Political cartoonists kept track of his dealings and alliances, depicting him, in one illustration, as a colleague of J. Edgar Hoover, Attorney General Tom C. Clark, and W. Averell Harriman.[7] Nourse emerged from his tenure on Truman's council as one of the most influential agricultural economists of the twentieth century.

Yet while Nourse's conception prevails in American agriculture, Bailey's still survives. We find him in the work of writers like Wendell Berry and Gene Logsdon; in the historical writing of David Danbom, Colin Duncan, and Donald Worster; and in the researches of Wes Jackson, whose work to design agricultural ecosystems sustaining of human communities and whose conviction that science can serve ends other than profit and power is reminiscent of Bailey himself.

We find elements of Bailey's philosophy among students of agroecology and their analysis of single-crop cultivation. Monoculture, this book has argued, forced the formation of each of the key institutions that emerged in California agriculture, and it is monoculture that takes the sharpest barbs from students of sustainable farming.[8] Genetic uniformity not only causes out-of-control insect populations and labor shortages, it also raises the prospect of starvation. The erosion in the genetic diversity of crop plants and the narrowing of the species people depend upon to eat gambles with the sustenance of entire continents. Any virus or pest infestation that comes along might cause unforeseen shortages. In the words of Miguel A. Altieri: "Modern agriculture is shockingly dependent on a handful of varieties for its major crops." In the United States 60 to 70 percent of the bean crop consists of two to three varieties, 72 percent of the potato crop is made up of four varieties. And when the affluent countries export their bioengineered plants to the poor countries, genetic erosion spreads too. In Sri Lanka farmers once tended close to two thousand varieties of rice. They now grow five. India once knew thirty thousand varieties of rice. That country now depends on ten.[9] The single crop is a worldwide practice. Biodiversity would require a reinvention of agriculture at least as profound as the transition from extensive to intensive cultivation that took place in the first two decades of the twentieth century.[10]

It would also require a new mission for the University Experiment Stations. No one has detailed the false loyalties and failed duties of the land-grant colleges like Jim Hightower. In *Hard Tomatoes, Hard Times* Hightower argues that the land-grant colleges, originally established under directors like Eugene Hilgard to service farmers in their localities, now mostly represent those industrial operators capable of purchasing the expensive inputs that the stations invent. Writes Hightower, "The [college] complex has worked hand in hand with seed companies to develop high-yield seed strains, but it has not noticed that rural America is yielding up practically all of its young people." The colleges make themselves available to help nonfarming companies move into agriculture, "while offering independent, family farmers little more comfort than 'adapt or die.'"[11] Hightower calls on the colleges to return to their care for local needs and their respect for farmers on small acres.

What unifies these thinkers is a resistance to "commercial valuations of life" and the conviction that rural places have a value apart from their capacity to produce export commodities. They recognize, moreover, that no other aspect of our relationship with nature is as important as agriculture. Farming remains what it has always been: the central biological and ecological relationship in any settled society, the source of our material production and human reproduction, and the most profound way humankind has changed the earth.[12] Though environmentalists have long insisted that in the unplowed, unimproved places lies the preservation of the world, that view neglects the most obvious source of our preservation. It lies, instead, in the nature we have mixed up with ourselves, in the cultivated lands, in the middle landscape that we can still reclaim.[13]

Reclaiming begins with understanding. As Raymond Williams reminds us, "The common image of the country is now an image of the past, and the common image of the city an image of the future. That leaves, if we isolate them, an undefined present."[14] This book has been an attempt to define the countryside of the present by way of the past. Seeing the countryside as a place in the present might lead Americans to pay closer attention to their food and the systems that sustain them. It might also broaden environmentalism to include the countryside, its contaminated soil and water, and the poverty of many rural people. We can refuse nostalgia and Golden Ages and still recognize that farmers of the distant past have something to teach; we can refuse modernist fantasies and still use science to humane ends. We can resolve to mend the countryside as one of many other ways to mend the world.

Notes

Preface

1. Mary Eschelbach Gregson, "Specialization in Late-Nineteenth-Century Midwestern Agriculture: Missouri as a Test Case," *Agricultural History* 67 (winter 1993): 16–36.

2. Colin A. M. Duncan, *The Centrality of Agriculture: Between Humankind and the Rest of Nature* (Montreal and Kingston: McGill-Queen's University Press, 1996).

Prologue: In the Rain's Shadow

1. French Strother, "Building a Wonderful Community: How a Sandy Waste Became a Prosperous Agricultural Region," *World's Work* 9 (February 1905): 5826–41; Clarence E. Edwards, "Special Crops of the Pacific Coast," *American Review of Reviews* 40 (July 1909): 64.

2. Edwards, "Special Crops," 63.

3. Garci Rodriquez Ordonez de Montalvo, "The California of Queen Calafia," in John Caughey and LaRee Caughey, eds., *California Heritage: An Anthology of History and Literature*, rev. ed. (Itasca, Ill.: F. E. Peacock, Publishers, 1971), 49.

4. James J. Parsons, "The Uniqueness of California," *American Quarterly* 7 (spring 1955): 45–55.

5. Carey McWilliams, *Southern California Country* (New York: Duell, Sloan, and Pearce, 1946), 6–7, quoted in Parsons, "The Uniqueness of California," 48. McWilliams was referring to the southern part of the state but the same holds for California as a whole.

6. David Hornbeck and Phillip Kane, eds., David L. Fuller, cartography, *California Patterns: A Geographical and Historical Atlas* (Mountain View, Calif.: Mayfield Publishing, 1983), 20.

7. Ibid., 21.

8. There were other crop-producing areas: the coast north of San Francisco, with its sheep economy, and the Imperial Valley on the Mexican border. The

latter became an important producer of vegetables, figs, dates, and flowers, all irrigated by the Colorado River. Although events in the northern part of the state
do not differ significantly from the development of Imperial Valley, this study
does not consider it in any detail. For a narrative that does, see Donald Worster,
Rivers of Empire: Water, Aridity, and the Growth of the American West (New York:
Pantheon, 1985).

9. Otto Hackel, "Summary of the Geology of the Great Valley," in Edgar H.
Bailey, ed., *Geology of Northern California*, Bulletin 190 of the California Divi-
sion of Mines and Geology (San Francisco: California State Division of Mines and
Geology, 1966), 217.

10. Tributaries to the San Joaquin River include (moving south to north)
the Kings, Merced, Tuolumne, and Calavaras Rivers. See David Hornbeck and
Phillip Kane, *California Patterns*, 13.

11. This figure is a rough estimate based on the following: California had
31 percent of the fruit acreage in the United States and harvested 36 percent of all
the land planted in fruit in 1929. United States Department of Commerce, Bureau of the Census, *Fifteenth Census of the United States: 1930—Agriculture, Volume III, Type of Farm, Part 3—The Western States* (Washington, D.C.: Government Printing Office, 1932), 12–17; United States Department of Agriculture,
Agriculture Yearbook, 1925 (Washington, D.C.: Government Printing Office,
1926), 880. The state produced between 60 and 100 percent of key fruit crops.
United States Department of Agriculture, *Yearbook of Agriculture: 1930* (Washington D.C.: Government Printing Office, 1930), 726–47.

12. "All land in farms" includes cotton, dairy, stock, and ranch uses, and land
supporting poultry. In 1930 the total land in farms in the state of California was
30,442,581 acres. U.S. Department of Commerce, Bureau of the Census, *Fifteenth Census of the United States: 1930—Agriculture, Volume III*, 12–17; California Development Association, *Report on Problems of Agricultural Development in
California, Prepared by the Department of Research and Information* (San Francisco: California Development Association, 1924).

13. The percentage refers to farm value. The total farm value of California agriculture in 1929 reached $623,103,467, to which fruit contributed $218,045,083.
United States Department of Commerce, Bureau of the Census, *Fifteenth Census of the United States: 1930—Agriculture, Volume III*, 396; California Development Association, *Report on Problems of Agricultural Development*.

14. United States Department of Agriculture, *Yearbook of Agriculture: 1930*,
726–47. Also see Peveril Meigs, "Current Trends in California Orchards and
Vineyards," *Economic Geography* 17 (July 1941): 275.

15. Los Angeles, San Bernardino, Tulare, and Ventura Counties produced
most of the citrus in the United States. California produced 18,537 cases out of a
total of 27,564. United States Department of Agriculture, *Yearbook of Agriculture: 1930*, 726–47.

16. United States Department of Agriculture, *Agriculture Yearbook, 1925*, 880.
The *Yearbook* of 1925 is dedicated to commercial fruit crops. San Diego, San Bernardino, Los Angeles, San Luis Obispo, San Benito, and Merced Counties also
grew peaches. Cleland and Hardy, *March of Industry*, appendix, maps.

17. United States Department of Agriculture, *Agriculture Yearbook: 1930*,
974–75.

1. The Conservation of the Countryside

1. For an account of public anxiety over the perceived closing of the frontier see David M. Wrobel, *The End of American Exceptionalism: Frontier Anxiety from the Old West to the New Deal* (Lawrence: University Press of Kansas, 1993).

2. The "Golden Age" of agriculture refers to the years between 1909 and 1920, when the prices of farm commodities and city-made goods attained a brief parity. It came to an end with the farm crisis that hit with the end of World War I and the subsequent collapse in commodity prices. See David Friday, "The Course of Agricultural Income during the Last Twenty-Five Years," *American Economic Review* 13 (March 1923): supplement; and James Shideler, *Farm Crisis: 1919–1923* (Berkeley and Los Angeles: University of California Press, 1957).

3. Edwin G. Nourse, "The Revolution in Farming," *Yale Review*, n.s., 8 (October 1918): 93.

4. "Physical Output and Farm Price of Individual Products, 1897–1939," in Harold Barger and Hans H. Landsberg, *American Agriculture, 1899–1939: A Study of Output, Employment, and Productivity* (New York: National Bureau of Economic Research, 1942), 332.

5. United States Department of Commerce, Bureau of the Census, *Twelfth Census of the United States, Agriculture*, vol. VI, pt. II, "Crops and Irrigation," (Washington, D.C.: Government Printing Office, 1902), 15; United States Department of Commerce, Bureau of the Census, *Thirteenth Census of the United States, Agriculture, General Report* (Washington, D.C.: Government Printing Office, 1913), 535–65. Another statistic that sheds light on the progress of agriculture, per capita value, rose from $40 per person in 1900 to $60 in 1910, a 51.2 percent increase. The value of all agricultural products rose by 83 percent in the same period. Ibid., 535.

6. Friday, "Agricultural Income," 152.

7. Sherman Johnson, "Corn: Acreage and Production in the United States, 1910–45" and "Farm Output per Capita, and Total Cropland Per Capita, 1919–44," in *Changes in Farming in War and Peace* (Washington, D.C.: United States Bureau of Agricultural Economics, 1946). Barger and Landsberg compiled an index of agricultural output in 1942. They set output in 1899 equal to 100 and listed the basic index as well as a five-year moving average, up to the year 1939. Between 1897 and 1920 the index moved from 97 to 130. In 1906 it stood at 118 points, fell to 110 in 1907, then recovered to 112 the next year. Over the forty-year period displayed by the table, farm output grew by about 1 percent a year. Barger and Landsberg, *American Agriculture*, 21, 53.

8. *Twelfth Census of the United States, Agriculture*, vol. V, pt. I, xvii; *Thirteenth Census of the United States, Agriculture*, vol. I, 28; *Fourteenth Census of the United States, Agriculture* (Washington D.C.: Government Printing Office, 1920), 24. Cities went from about 42 million to about 54 million people during the second decade of the twentieth century, for a growth rate of 28.8 percent. By 1920 the population of the United States was more than 50 percent urban (48.67 percent rural).

9. Edward F. Adams, "The Position of the Farmer in Our Economic Society," in Edwin G. Nourse, ed., *Agricultural Economics* (Chicago: University of Chicago Press, 1916), 75; A. C. True observed: "Here is the real issue: On the

success of the 30,000,000 people on our 6,000,000 farms, who do not produce an annual wealth of nine billion dollars, depend the life, comfort, and prosperity of the 60,000,000 people engaged in trade, manufacturing, transportation, the professions and government." A. C. True, "Some Fundamentals of Agricultural Progress," *Transactions of the Commonwealth Club of California* 6 (11 October 1911): 446.

10. *New York Times*, 6 July 1908.

11. Ibid., 19 July 1908.

12. Friday, "Agricultural Income," 152.

13. *New York Times*, "Global Industrial Crisis," 20 May 1907; Ibid., "Farmers Dependent upon Foreign Markets," 1 January 1907; ibid., "Commodity Prices Explained," 20 January 1907; ibid., 24 May 1907; John G. Thompson, "The Nature of Demand for Agricultural Products and Some Important Consequences," *Journal of Political Economy* 24 (February 1916): 173–76.

14. Liberty Hyde Bailey, *The Country-Life Movement in the United States* (1911; reprint, New York: Macmillan, 1915), 34.

15. *Report of the Country Life Commission*, United States Senate Document No. 705, 60th Cong., 2nd Sess. (Washington, D.C.: Government Printing Office, 1909), 17. The commission: Liberty Hyde Bailey, Henry Wallace, Kenyon L. Butterfield, Walter H. Page, Gifford Pinchot, C. S. Barrett, W. A. Beard.

16. Macy Campbell, *Rural Life at the Crossroads* (Boston: Ginn and Company, 1927), 1.

17. Ibid., 24.

18. Theodore Roosevelt to William A. Beard, 11 November 1908, Theodore Roosevelt Papers, series 2:322–26, microfilm, Sterling Library, Yale University.

19. Ibid.

20. Liberty Hyde Bailey, *The Outlook to Nature* (New York: Macmillan, 1905), 82.

21. Ibid., 108–9.

22. *New York Times*, 18 July 1907.

23. Bailey's major intellectual innovation was to apply the principles of botanical science to the study of flowers, fruits, and other garden plants, thus establishing horticulture as a major science. His many books, handbooks, and especially his *Cyclopedia of Horticulture* defined subtle points for thousands of farmers and fruit growers.

The view of Bailey presented here conflicts with that presented by David Danbom in *The Resisted Revolution: Urban America and the Industrialization of Agriculture, 1900–1930* (Ames, Iowa: Iowa State University Press, 1979). Danbom defines Bailey as an "urban agrarian" who "directed [his] appeal to urban people uneasy about the city and life therein." He groups Bailey with those reformers who turned to the countryside "for solutions to urban problems." While Bailey addressed urban audiences at different times, the vast body of his work sought answers to distinctly rural problems.

24. There are two biographies of Liberty Hyde Bailey: Philip Dorf, *Liberty Hyde Bailey: An Informal Biography* (Ithaca, N.Y.: Cornell University Press, 1956) and Andrew Denny Rodgers III, *Liberty Hyde Bailey: A Story of American Plant Sciences* (Princeton, N.J.: Princeton University Press, 1949).

25. Liberty Hyde Bailey, *The Harvest of the Year to the Tiller of the Soil* (New York: Macmillan, 1927), 27.

26. Ibid., 28.

27. Bailey, *Harvest of the Year*, 29–30.

28. Bailey, *The Holy Earth* (Ithaca, N.Y.: Comstock Publishing, 1919), 24.

29. Ibid., 30; Bailey, *Harvest of the Year*, 107–8 and *Country-Life Movement*, 56–57.

30. Liberty Hyde Bailey, "The Collapse of Freak Farming," *Country Life in America* 4 (May 1903): 14.

31. *Report of the Country Life Commission*, 57–58.

32. Joseph G. Knapp, *Edwin G. Nourse—Economist for the People* (Danville, Ill.: Interstate Printers and Publishers, 1979), 31–46. It is not clear whether Nourse and Bailey knew each other.

33. E. G. Nourse, "The Aim and Scope of Agricultural Economics," introduction to *Agricultural Economics*, ed. by E. G. Nourse, 9.

34. E. G. Nourse, "The Place of Agriculture in Modern Industrial Society," *Journal of Political Economy*, 27 (July 1919): 561–77, reprinted as "Agriculture and Modern Industry," in Louis Bernard Schmidt and Earle Dudley Ross, eds., *Readings in the Economic History of American Agriculture* (New York: Macmillan, 1925), 581–82.

35. Nourse, "Revolution in Farming," 90–91.

36. Nourse, "Place of Agriculture," 577.

37. Ibid.

38. Leo Marx, *The Machine in the Garden: Technology and the Pastoral Ideal in America* (New York: Oxford University Press, 1964), 150–69, 181.

39. The decline of the People's Party signified the end of the countryside as an independent political entity. This is the view of Lawrence Goodwyn, *Democratic Promise: The Populist Moment in American History* (New York: Oxford University Press, 1976). Also see Martin Ridge, *Ignatius Donnelly: The Portrait of a Politician* (Chicago: University of Chicago Press, 1962). For a general bibliography see Richard L. McCormick, *Public Life in Industrial America, 1877–1917*, part of The New American History, a series edited by Eric Foner and published by the American Historical Association.

40. Bailey criticized efficiency as the paramount value of the times: "The idea of mechanical efficiency has so far invaded our lives that many persons would introduce it into the home and even into play; yet our very need in a machine age is to encourage informality and animation, that is, personality, rather than make a person an automaton or mechanical toy." Bailey, *Harvest of the Year*, 107–8; Bailey, *Country-Life Movement*, 56–57; for a discussion of efficiency in America during the progressive era see Samuel Haber, *Efficiency and Uplift: Scientific Management in the Progressive Era, 1890–1920* (Chicago: University of Chicago Press, 1964).

41. David Danbom argues that the various legislative acts and the general program for educational and institutional reform that comprised the Country-Life Movement met widespread resistance among rural people. His book is convincing on many points, especially about the mood in American cities regarding the health of the countryside. Yet it is difficult to rectify the resistance he details

with the push for greater output that, Danbom admits, farmers expressed. Danbom, *Resisted Revolution*, 90, 133. Kenyon Butterfield, *Chapters in Rural Progress* (Chicago: University of Chicago Press, 1907); Edward F. Adams, *The Modern Farmer in His Business Relations* (San Francisco: N. J. Stone, 1899). For a description of how farmers reacted to experiment station scientists and technology see Deborah Fitzgerald, *The Business of Breeding: Hybrid Corn in Illinois, 1890–1940* (Ithaca: Cornell University Press, 1990).

42. James J. Hill, *Address Delivered by Mr. James J. Hill before the Farmers' National Congress, Madison, Wisconsin, September 24, 1908*, published separately, but collected in one volume by the Huntington Library, San Marino, California.

43. Richard White, *"It's Your Misfortune and None of My Own," A History of the American West* (Norman, Okla.: University of Oklahoma Press, 1991), 433; Donald Worster writes that after 1912, when Congress reduced the time for improvements before the final transfer of title from five to three years, there were twenty-four thousand land entries in the West and fifty-three thousand the next year. "And they remained over 30,000 annually until the early twenties." Worster, *Dust Bowl: The Southern Plains in the 1930s* (Oxford: Oxford University Press, 1979), 88.

44. In 1820, "80 per cent of the food products of northern farms were consumed by the rural population"; the rest went to local urban populations in the northern states and abroad. "By 1870 the portion of farm production consumed within the rural community had fallen to about 60 per cent." Clarence H. Danhof, *Change in Agriculture: The Northern States, 1820–1870* (Cambridge, Mass.: Harvard University Press, 1969), 9–26.

45. Differences in grades and telegraphic communication eventually made it so that farmers of the same commodities all over the country took close to the same price. See Cronon, *Nature's Metropolis*, 109–21.

46. John Faragher writes that the people who traveled overland from eastern states to the Pacific Coast sought fresh land: "Over a quarter of the writers of the diaries and recollections stated unequivocally that new agricultural land was *the* motive in their decision to emigrate." John Mack Faragher, *Women and Men on the Overland Trail* (New Haven: Yale University Press, 1979), 16; Clarence Danhof, *Change in Agriculture*, 150.

47. Donald Worster, *Dust Bowl*, 84.

48. E. C. Chilcott, "Dry-Land Farming in the Great Plains Area," *Yearbook of the United States Department of Agriculture, 1907* (Washington, D.C.: Government Printing Office, 1908), 451–68.

49. For works on Great Plains history and environment see Mary W. M. Hargreaves, *Dry Farming in the Northern Great Plains: Years of Readjustment, 1920–1990* (Lawrence: University Press of Kansas, 1993); John Ise, *Sod and Stubble: The Story of a Kansas Homestead* (Lincoln: University of Nebraska Press, 1936); James C. Malin, *The Grasslands of North America: Prolegomena to its History* (Ann Arbor, Mich.: James C. Malin, 1948); Rodman W. Paul, *The Far West and the Great Plains in Transition, 1859–1900* (New York: Harper and Row, 1988); Wallace Stegner, *Beyond the Hundredth Meridian: John Wesley Powell and the Second Opening of the West* (Lincoln: University of Nebraska Press, 1982); Donald Worster, *Dust Bowl*; Walter Prescott Webb, *The Great Plains* (1931; reprint, Lincoln: University of Nebraska Press, 1981).

50. T. D. Rice and J. A. Warren, "Relation of Agricultural Industries to Natural Environment in the Great Plains," in *Source Book for the Economic Geography of North America*, edited by Charles C. Colby, 3rd ed. (Chicago: University of Chicago Press, 1926), 342.

51. Hargreaves, *Dry Farming*, 4. Donald Worster described the "new-style sodbuster" of the twentieth century as "an expansionist, feeling all the old land hunger of an opportunity-seeking democrat, but adding an intense desire to make his new machines profitable that would have shocked Thomas Jefferson's agrarian idealism." Worster, *Dust Bowl*, 87–88

52. Rudolf Cronau, *Our Wasteful Nation: The Story of American Prodigality and the Abuse of our Natural Resources* (New York: M. Kennerley, 1908).

53. James J. Hill, *Address*, 2.

54. Milton M. Snodgrass and Luther T. Wallace, *Agriculture, Economics, and Growth*, 2nd ed. (New York: Appleton-Century-Croft, 1970), 228.

55. Willard W. Cochrane, *The Development of American Agriculture: A Historical Analysis*, 2nd ed. (Minneapolis: University of Minnesota Press, 1993), 342.

56. Hugh M. LaRue, "Opening Address," *Transactions of the California State Agricultural Society* (Sacramento: State Printing Office, 1883), 18.

57. The classic discussion of how farmers and crops built an important agricultural region is Allan G. Bogue's, *From Prairie to Corn Belt: Farming on the Illinois and Iowa Prairies in the Nineteenth Century* (Chicago: University of Chicago Press, 1963; Chicago: Quadrangle Paperback, 1968), 123–47. For other studies on the same subject: William Cronon, *Nature's Metropolis;* John Mack Faragher, *Sugar Creek: Life on the Illinois Prairie* (New Haven: Yale University Press, 1986); Gilbert C. Fite, *The Farmers' Frontier, 1865–1900* (New York: Holt, Rinehart, and Winston, 1966). On agricultural specialization see Clarence Danhof, *Change in Agriculture*, 9–26. For more on farming in the United States see Percy Bidwell and John Falconer, *History of Agriculture in the Northern United States* (Washington, D.C.: Carnegie Institution, 1925); Paul W. Gates, *The Farmer's Age: Agriculture 1815–1860* (New York: Holt, Rinehart, and Winston, 1960); Fred Shannon, *The Farmer's Last Frontier*. Also see Edwin G. Nourse, *Agricultural Economics;* and Mary Eschelbach Gregson, "Specialization in Late-Nineteenth-Century Midwestern Agriculture," 16–36.

58. The collective body of books and articles written on the subject of regional specialization might be called the "land utilization school" of American economic geography. Works in this field include detailed studies as well as grade school textbooks—all intended to inventory the agricultural wealth of the United States. Major works include Oliver Edwin Baker, "The Agricultural Regions of North America," published in installments in *Èconomic Geography* vol. 2 (October 1926); vol. 3 (January, July, October 1927); vol. 4 (January, October 1928); vol. 5 (January 1929); and vol. 6 (April 1930); Baker, "The Agriculture of the Great Plains Region," *Annals of the Association of American Geographers* 13 (September 1923); Baker and H. M. Strong, "Arable Lands in the United States," *United States Department of Agriculture Yearbook, 1918* (Washington, D.C.: Government Printing Office, 1919), 433–41; Baker, ed., *Atlas of American Agriculture* (Washington, D.C.: Government Printing Office, 1918); Baker, ed., *A Graphic Summary of American Agriculture, Based Largely on the Census of 1920, USDA Yearbook, 1921* (Washington, D.C.: Government Printing Office, 1921); Frederick Anthony

Buechel, *The Commerce of Agriculture: A Survey of Agricultural Resources* (New York: John Wiley and Sons, 1926); Charles C. Colby, ed., *Source Book for the Economic Geography of North America*, 3rd. ed. (Chicago: University of Chicago Press, 1926); Richard Elwood Dodge, "Some Geographic Relations Illustrated in the Practice of Agriculture," *Bulletin of the American Geographic Society* 44 (April 1912); and Joseph Russell Smith, *North America; Its People and the Resources, Development, and Prospects of the Continent as an Agricultural, Industrial, and Commercial Area* (New York: Harcourt, Brace and Company, 1925); as well as the many publications of L. C. Gray—for instance, Gray, "The Utilization of Our Lands," *Agricultural Yearbook, 1923* (Washington D.C.: Government Printing Office, 1924). Frederick Jackson Turner offers the section, a politically defined and more broadly conceived kind of region, as the unit most significant in understanding the West after its initial period of settlement. Turner interprets the regional specialization of farming as an attempt to stop Eastern money and power from rolling over the West. The natural attributes "of certain regions for farming . . . will arrest the tendency of the Eastern industrial type of society to flow across the continent and thus to produce a consolidated, homogeneous nation free from sections." Turner, "The Significance of the Section in American History," in *The Significance of Sections in American History* (New York: Henry Holt and Company, 1932), 35.

59. Baker's dates: 1883–1949. See Richard Kirkendall's entry on Baker in John A. Garraty and Edward T. James, eds., *The Dictionary of American Biography*, supplement 4, 1946–50 (New York: Charles Scribner's Sons, 1974), 44–46. See also Kirkendall, "L. C. Gray and the Supply of Agricultural Land," *Agricultural History* 37 (October 1963): 206–14. Baker's work represents the empirical wing of location theory; that is, he did not construct models like Von Thünen but drew conclusions about how the world worked by observing it. For a summary of location theory and its contemporary applications see Leslie Symons, *Agricultural Geography* (New York: Frederick A. Praeger, 1967).

60. In this case, "competition" does not mean a contest for the buyer by offering the choicest produce (although quality mattered) but a contest on the production side for the lowest costs and thus the highest profits. Since produce in the large urban markets received similar prices, only the farmers who produced at the lowest cost made money and stayed in business.

61. Johann Heinrich Von Thünen, the German economist, described this situation when he wrote, "When such improvements are bought at a cost higher than the value of the additional product they achieve, they not only ruin the farmer who undertakes them, but reduce the total national wealth." Johann Heinrich Von Thünen, *Der Isolierte Staat*. In *Von Thünen's Isolated State: An English Translation of Der Isolierte Staat*, ed. by Peter Hall and trans. by Carla M. Wartenberg. (Oxford, England: Pergamon, 1966), 2:231.

62. Oliver E. Baker, "The Increasing Importance of the Physical Conditions in Determining the Utilization of Land for Agriculture and Forest Production in the United States," *Annals of the Association of American Geographers* 11 (1921): 45. Richard Kirkendall writes that Baker served as L. C. Gray's "top lieutenant" during the 1920s. Gray said many of the same things and encouraged a public policy of "deliberate selection, careful economy, and constructive development," ideas quite at odds with the religion of expansion. Gray opposed reclamation

boosters and extensive agriculture. See Kirkendall, "L. C. Gray and the Supply of Agricultural Land," 207–9.

63. Baker, "Physical Conditions," 45.

64. David Ricardo, *On the Principles of Political Economy and Taxation*, published as part of *The Works and Correspondence of David Ricardo*, edited by Piero Sraffa, with the collaboration of M. H. Dobb, vol. I (Cambridge: Cambridge University Press, 1951; reprint, 1975), 67; Robert Heilbroner, *The Worldly Philosophers: The Lives, Times and Ideas of the Great Economic Thinkers* (New York: Simon and Schuster, 1953; reprint, 1986), 96–99.

65. This is the origin of Ricardo's phrase: "Corn is not high [in price] because a rent is paid, but a rent is paid because corn is high [in price]." By adjusting for the fact that Ricardo equated cost of production with price, we can understand him to mean that corn raised on inferior land is expensive to produce, not because the farmer pays rent; but rather, the farmer pays rent on inferior land because corn is expensive to produce there. Ibid., 77.

66. Ibid., 74–80.

67. Ibid.; Heilbroner, *Worldly Philosophers*, 96–99; Baker, "Physical Conditions," passim.

68. *The Isolated State* touches on taxation, land rent, methods of animal husbandry and agriculture, forestry, and intensive farming. Von Thünen, in *Von Thünen's Isolated State*, 9. Models of the economic world can only explain so much. As Leslie Symons wrote about Thünen and other world builders, "No model can yet cope with the complexity of background, the intricacy of human reasoning and the constant change that affects all economies." Symons, *Agricultural Geography*, 206. Part I of the *Isolated State*, the section dealing with agriculture and crop zones, first appeared in Hamburg in 1826; ibid., xviii. For a remarkable account of Von Thünen's theory as it applies to nineteenth-century Chicago see Cronon, *Nature's Metropolis*.

69. Cronon, *Nature's Metropolis*, 97.

70. G. F. Warren, *Farm Management* (New York: Macmillan, 1919), 51–52.

71. Ibid.

72. Ibid., 39.

73. Baker, "Physical Conditions," 39. My emphasis.

74. William H. Mills, "Annual Address," *Transactions of the California State Agricultural Society* (Sacramento: State Printing Office, 1890), 188–89.

75. Ibid., 193.

76. William H. Brewer, *Up and Down California in 1860–1864*, edited by Francis P. Farquhar, rev. ed. (Berkeley and Los Angeles: University of California Press, 1974), 202–3, 382.

77. Thompson, *Official Atlas Map of Fresno County*, 8–9; William S. Chapman, letter of 27 August 1868 in the *San Francisco Bulletin*, published in Gerald D. Nash, "Henry George Reexamined: William S. Chapman's Views on Land Speculation in Nineteenth Century California," *Agricultural History* 33 (1959): 133–37.

78. Charles Nordhoff, *California for Health, Pleasure, and Residence: A Book for Travellers and Settlers* (1876; reprint, Ten Speed Press, 1973), 184.

79. As early as 1860 almost one-half of the wheat and flour exported from California went to British consignors. The rest went to Australia, New York, and China. For a brief history of the wheat trade and its players see Rodman Wilson

Paul, "Great California Grain War: The Grangers Challenge the Wheat King," *Pacific Historical Review* 27 (November 1958); and Paul "The Wheat Trade between California and the United Kingdom," *Mississippi Valley Historical Review* 45 (December 1958): 391.

80. Rodman Paul quotes the former president of the San Joaquin Valley Agricultural Society who stated that the people of rural California "were better suited to 'managing general business' than to 'tilling the soil.'" Letter from E. S. Holden, 21 February 1871 to the *Pacific Rural Press*, 11 March 1871, quoted in Rodman W. Paul, *Far West;* Paul, "Wheat Trade," 391.

81. Paul W. Gates, "Public Land Disposal in California," *Agricultural History* 49 (January 1975); Arthur Maass and Raymond L. Anderson, . . . *And the Desert Shall Rejoice: Conflict, Growth, and Justice in Arid Environments* (Cambridge: MIT Press, 1978), 158–59. The Swamp Land Act (1855 and often amended) was an act of the California State Legislature.

82. Gates, "Public Land Disposal," 175; Arthur Maass and Raymond L. Anderson, . . . *And the Desert Shall Rejoice,* 157; Donald J. Pisani, *From Family Farm to Agribusiness: The Irrigation Crusade in California and the West, 1850–1931* (Berkeley and Los Angeles: University of California Press, 1984), 15. Also see Henry George, "Our Land and Land Policy," vol. 9 of *Writings of Henry George* (New York: Doubleday and McClure, 1901), 71.

83. William S. Chapman took credit for this observation in his published letter of 27 August 1868, in which he wrote, "Where such grass would grow as there grew in winter and withered in summer, wheat would grow. But I thought that the wheat ought to have the same time to grow as the grass: That it ought to be sowed just before or at the very commencement of the wet season. That it might grow stout and thick so as to cover the roots completely from the burning rays of the summer sun, when the rain should cease." William S. Chapman, published in Nash, "Henry George Reexamined," 136.

84. The figures are for a wheat field of 7,330 acres, yielding eleven bushels per acre. The total cost of harvesting came to $5,913 or $.80 per acre. Figures compiled by Frank T. Latta from the records of the wheat farmer and inventor L. A. Richards of Grayson, Stanislaus County [c. 1876], in Latta Collection, Sky-Farming Papers, folder 1(10)1, Huntington Library Manuscript Collection; *Pacific Rural Press,* 4 January 1879, 1 February 1879, 8 September 1883; Joseph Hutchinson, "California Cereals.—II," *Overland Monthly,* 2nd ser. (August 1883); Richard Hilkert and Oscar Lewis, *Breadbasket of the World: California's Great Wheat Growing Era: 1860–1890* (San Francisco: Book Club of California, 1984).

85. Albert T. Webster, "A California Wheat Harvest," *Appleton's Journal,* [1876?], 457, Beinecke Rare Book and Manuscript Library, Yale University.

86. Ibid., 452, 453. This wheat harvest called for nine hundred men and one thousand horses. On Hugh Glenn's ranch in Colusa County the harvest amounted to five hundred thousand bushels one year and required 1,250 railroad cars to haul it away. "Hugh Glenn," in Hilkert and Lewis, *Breadbasket of the World.*

87. Gates, "Public Land Disposal in California"; Pisani, *Family Farm to Agribusiness,* 119–27. For other fortunes see Frank F. Latta, ed., "The Story of Henry Hammer Mills," unpublished oral history in the Sky Farming papers,

Frank F. Latta Collection, The Huntington Library; Khaled Bloom, "Pioneer Land Speculation in California's San Joaquin Valley," *Agricultural History* 57 (July 1983): 305.

88. Henry George, "Our Land and Land Policy," 9:68.

89. Thompson, *Official Atlas Map of Fresno County*, 13. The same six sections planted in fruit would gross between $500,000 and $700,000 a year, netting each twenty-acre farm between $2,500 and $3,500 a year, so estimated the author.

90. Nordhoff, *California*, 130–31, 187.

91. William H. Mills, "Annual Address," 188.

92. Charles F. Reed, "What Is Wanted in California," *California Horticulturist and Floral Magazine*, 1 (November 1870): 8. By 1893, a bushel sold for $.53, and the product itself declined in quality. Tired land made for gluten-poor flour and dissatisfied English bakers. Donald Pisani, *From Family Farm*, 288–90. The hundred-pound bag, or cental, was a unit used almost exclusively in the British trade.

93. Wheat lands experienced a widespread loss in fertility that brought yields below ten bushels per acre in many regions. Even back in 1873, after only a few years of cultivation, farmers took only ten to fifteen bushels from an acre in the San Joaquin. Nordhoff, *California*, 185; "Irrigation in the Great Central Valley," *California Horticulturist and Floral Magazine* 3 (September 1873): 280.

94. *Pacific Rural Press*, 20 (2 October 1880).

95. Ibid.

96. N. A. Cobb, *The California Wheat Industry*, Miscellaneous Publication No. 159 (Sydney, New South Wales: Department of Agriculture, 1901), 25.

97. According to Rodman Paul, shipping rates actually fell in the 1880s, when larger sailing ships entered the trade. The wheat boom might have ceased earlier than it did had not the rates dropped. Paul, "Wheat Trade," 410–11.

98. Chapman letter in Nash, "Henry George Reexamined," 133–37. The letter is a forceful apologia for speculation, in which Chapman writes: "Men value little what costs them little or nothing. . . . Just so, as long as the land in the San Joaquin Valley could be had for the asking, nobody wanted it." Ibid., 136; Nordhoff, *California*, 128.

99. Donald Pisani calls irrigation during this time "the stepchild of land speculation." See Donald Pisani, *From Family Farm*, 79; Robert Hine, *California's Utopian Colonies* (Berkeley and Los Angeles: University of California Press, 1954; reprint, New York: W. W. Norton and Company, 1969).

100. William E. Smythe, *The Conquest of Arid America* (New York: Harper and Brothers Publishers, 1900), 132–36; E. J. Wickson, "1920 State Fair Annual," 27, in E. J. Wickson Papers, Bancroft Library.

2.   Orchard Capitalists

1. For literature on farming as a business see Clarence Danhof, *Change in Agriculture;* Adam Ward Rome, "American Farmers as Entrepreneurs, 1870–1900," *Agricultural History* 56 (January 1982); Allan Bogue, *From Prairie to Cornbelt;* Rodman W. Paul, "The Beginnings of Agriculture in California: Innovation vs.

Continuity," *California Historical Quarterly* 52 (1973); Howard Seftel, "Government Regulation and the Rise of the California Fruit Industry: The Entrepreneurial Attack on Fruit Pests, 1880–1920," *Business History Review* 59 (autumn 1985).

2. Howard Seftel makes the same point in "Government Regulation," 375.

3. Timothy Paige, *Farming That Pays* (San Francisco: n.p., 1891).

4. No other subject in the historiography of California has been as well documented as irrigation. For a range of recent interpretations see Donald Pisani, *From Family Farm* and *To Reclaim a Divided West: Water, Law, and Public Policy, 1848–1902* (Albuquerque: University of New Mexico Press, 1992); Norris Hundley Jr., *Water and the West: The Colorado River Compact and the Politics of Water in the American West* (Berkeley and Los Angeles: University of California Press, 1975) and *The Great Thirst: Californians and Water, 1770s–1990s* (Berkeley and Los Angeles: University of California Press, 1992); Donald Worster, *Rivers of Empire;* William L. Kahrl, ed., *The California Water Atlas* (Sacramento: Governor's Office of Planning and Research, 1979).

5. *Fresno Morning Republican,* 1 January 1896. For a more detailed discussion of land monopoly and economic development in the arid West, see Donald Pisani, *From Family Farm,* 124–28; Arthur Maass and Raymond L. Anderson, *. . . And the Desert Shall Rejoice,* 213 and passim. For a discussion of land development in a humid environment in which speculation played a positive role, see William Wyckoff, *The Developer's Frontier: The Making of the Western New York Landscape* (New Haven: Yale University Press, 1988).

6. Other colony builders used the same template for more elevated ends, founding plantations with spiritual and utopian missions. The Mormons of Utah integrated fields and waterways into the design of their towns. The members of the Greely Colony in Colorado used irrigation to found an isolated community based on cooperative ownership and a shared philosophy. But although developers sometimes attempted to attract like-minded people, they used the colony to settle as many families as possible on a single section and to build canals and ditches with the least amount of material. Developers sold to individuals more often than groups, but the term "colony" remained. John Wesley Powell concluded from the Mormon example that people would establish themselves in the arid regions only when they joined together to share the hardships of settlement. See Robert Hine, *California's Utopian Colonies.*

7. Thompson, *Atlas Map of Fresno County,* 12.

8. Charles Nordhoff, *California,* 175.

9. Maass and Anderson, *. . . And the Desert Shall Rejoice,* 157; Virginia Thickens, "Pioneer Agricultural Colonies of Fresno County," *California Historical Society Quarterly* 25 (March and June 1946): 175.

10. Thompson, *Atlas Map of Fresno County,* 12.

11. Joseph Russell Smith, *North America: Its People and the Resources, Development, and Prospects of the Continent as an Agricultural, Industrial, and Commercial Area* (New York: Harcourt, Brace and Company, 1925), 553; Lilbourne Alsip Winchell, *The History of Fresno County and the San Joaquin Valley* (Fresno: A. H. Cawston, 1933), 135.

12. P. Y. Baker, *The 76 Land and Water Company's Lands* (San Francisco: n.p., [1884?]), 20.

13. Pisani, *From Family Farm*, 122.

14. Frank Pixley, "An Idyl of the Tulare Lake Land," in *Secure a 20-Acre Home in the Tulare Colony* (San Francisco: Pacific Coast Land Bureau, 1885), 32. For other examples of the same see "General Directory of Fresno County, California, for 1881," 111–12, quoted in Maass and Anderson, . . . *And the Desert Shall Rejoice*, 167; Pisani, *From Family Farm*, 123.

15. Frank Pixley, "An Idyl of the Tulare Lake Land," 24.

16. George Clements, manager of the Agriculture Department at the Los Angeles Chamber of Commerce, said in 1920 that "the average orchardist is past forty-five years of age, has little, if any agricultural or horticultural knowledge, and if so blessed at all, it is not directly applicable to California conditions." George P. Clements, "Care of the Small Land Owner in Southern California," 11 June 1920, manuscript address, George Clements Papers, Bancroft Library.

17. P. Y. Baker, *76 Land and Water Company*. Speculation often inflated the price far beyond what the land developers quoted.

18. Gustav Eisen, *The Raisin Industry: A Practical Treatise on the Raisin Grapes, Their History, Culture, and Curing* (San Francisco: H. S. Crocker and Company, 1890), 179; C. F. Dowsett, *A Start in Life: A Journey across America. Fruit Farming in California* (London: C. F. Dowsett and Company, 1891), 83; Timothy Paige, *Farming That Pays*, 11. The census of 1890 listed $150 per acre for improved land. United States Department of Commerce, Bureau of the Census, *Eleventh Census of the United States—Agriculture* (Washington, D.C.: United States Census Office, 1890), 33.

19. Charles W. Clough and William B. Secrest, *Fresno County—the Pioneer Years, from the Beginnings to 1900* (Fresno: Panorama West Books, 1984), 143; Pisani, *From Family Farm*, 123. Their figures differ.

20. P. Y. Baker, *76 Land and Water Company*, 7; C. F. Dowsett, *A Start in Life*, 93. Both quote the same interest rate.

21. Section map of California, *The Garden of the World* (San Francisco: N. C. Carnall and Company, Dealers in Colony lands, [1890?]).

22. The cost of vines is figured at $.04 per vine and 544 vines per acre. Eisen, *Raisin Industry*, 179–80; Dowsett, *Start in Life*, 83–93; Paige, *Farming That Pays*, 11.

23. Albert Julius Winkler, *General Viticulture* (Berkeley and Los Angeles: University of California Press, 1974), 192, 612. In cooler regions vines will not bear at commercial levels for four years. The Thompson seedless will yield no grapes for the first two years. In the third it produces between one-sixth and one-third of a ton per acre. By the fifth it begins to "produce like mad"—up to twelve or thirteen tons per acre by the seventh year. Professor Dale Heine, department of agricultural economics, University of California at Davis, telephone interview by the author, 6 May 1991.

24. Richard T. Ely and George S. Wehrwein, *Land Economics* (Madison: University of Wisconsin Press, 1940; reprint, 1964), 257.

25. Dowsett, *Start in Life*, 93; Ely and Wehrwein, *Land Economics*, 257. Quoting data from the 1920s, Ely and Wehrwein recommend $500 for living expenses for the first year on an irrigated farm in Nevada. It was also common for new settlers to stay in their previous occupations while their vines or trees matured. Managers cared for the land in their absence.

26. This sum comes from adding the costs listed above, including the per-acre cost of cultivation over three years, minus what Dowsett states a family can earn with cover crops and eggs. Dowsett informed potential buyers that they could raise $650 for the first two years combined and $1,050 for the third. His figures, however, are for a forty-acre farm. Dividing in half, I came up with a gross profit of $850 for three years and subtracted that from my total cost estimate of $8,370. The sum of $7,500 is itself an estimate based on imperfect information and honest guesswork.

For comparison, Thomas Forsyth Hunt, dean of the College of Agriculture at the University of California, listed the average investment necessary to bring California farmland into production in 1910: under 20 acres, about $6,000; between 20 and 49 acres, $9,000; between 50 and 174 acres, $14,000; between 175 and 999 acres, $25,000; and over 1,000 acres, about $80,000. Thomas Forsyth Hunt, "Suggestions to the Settler in California," *University of California College of Agriculture Agricultural Experiment Station Circular No. 210* (March 1919): 4–5; Eisen, *Raisin Industry*, 179–80; Dowsett, *Start in Life*, 83–93.

27. Eisen, *Raisin Industry*, 179. Earl Pomeroy writes: "By 1916 the editor of the *Pacific Rural Press* estimated that under average conditions a farmer in California needed capital and credit of at least $20,000." Earl Pomeroy, *The Pacific Slope: A History of California, Oregon, Washington, Idaho, Utah, and Nevada* (New York: Knopf, 1965), 171.

28. *Fresno Irrigated Farms Co. Incorporated; Owners of the Celebrated Bank of California Tract . . . Facts for the Homeseeker* (San Francisco: Fresno Irrigated Farms Company, 1906), 47.

29. The promoter referred to tasks required during the first years of the orchard. *Fresno County, California—Where Can Be Found Climate, Soil, and Water, the Only Sure Combination for the Vineyardist* ([San Francisco]: Vogel, Hall, Lisenby), 1887.

30. Charles Carlson, unpublished diary, 30 September 1878, Charles Carlson Collection, Huntington Library.

31. Ibid., 4, February 1878.

32. Excerpts from diaries and additional notes by Carlson's daughter Alice Mary Carlson, photocopy of a typewritten manuscript, Carlson Collection, Huntington Library.

33. Charles Carlson, unpublished diary, 18 and 21 January 1882.

34. Carlson, diary, 1882.

35. William Smythe, *Conquest of Arid America*, 129.

36. Pisani, *From Family Farm*, 126.

37. *Fresno County, California, Its Offering for Settlement* (San Francisco: Pacific Coast Land Bureau, 1886), 13.

38. Paul E. Vandor, *History of Fresno County, California, with Biographical Sketches*, 2 vols. (Los Angeles: Historic Record Company, 1919), 363.

39. Gustav Eisen, "Why Some Raisin Vineyards Pay More Than Others," *California—A Journal of Rural Industry* 3 (22 March 1890).

40. Thomas Hill, *Irrigation at Strawberry Farm* (1865), Bancroft Library. Also see William H. Truettner, ed., *The West as America: Reinterpreting Images of the Frontier, 1820–1920* (Washington, D.C.: Smithsonian Institution Press, 1991), 230.

41. My thanks to Bill Cronon for this observation.

42. A. B. Butler, "California and Spain," *California—A Journal of Rural Industry* 3 (15 March 1890).

43. F. C. Barker, "The Gentleman-Farmer," *Irrigation Age* (May 1896): 205.

44. G. H. Hecke, "The Pacific Coast Labor Question from the Standpoint of a Horticulturist," in *Official Report of the Thirty-Third Fruit-Growers' Convention* (Sacramento: State Printing Office, 1908), 67–71.

45. L. A. Teague, "The Fresno Vineyards," *Overland Monthly* 52 (December 1908): 575. My emphasis.

46. Ibid.

47. One booster broadcast the success of a Fresno County raisin grower who had planted eighteen acres in vines "and has up to the present hired all the work done upon his place," C. A. Rohrabacher, *Fresno County, California: Descriptive, Statistical and Biographical* (Fresno: C. A. Rohrabacher, 1891), 25–26.

48. Charles Nordhoff, *California*, 133.

49. Rohrabacher, *Fresno County*, 29.

50. Ibid., 61–63; Thompson, *Atlas Map of Fresno County*, 13.

51. H. S. Drake, *Lindsay-Strathmore Land Development Directory* (Lindsay, Calif.: Hall and Burr, 1929).

52. "Who Grew the Raisins?" *California—A Journal of Rural Industry* 3 (12 April 1890); Eisen, "The Raisin Industry of Fresno," *California—A Journal of Rural Industry* 1 (February 1890).

53. "Woman Horticulturist," *California—A Journal of Rural Industry* 3 (24 May 1890).

54. Rohrabacher, *Fresno County*, 47–48.

55. Winfield J. Davis, *An Illustrated History of Sacramento County California* (Chicago: Lewis Publishing, 1890), 430–31.

56. *Who's Who in America*, vol. 14 (Chicago: A. N. Marquis, 1926), 923.

57. "Successful Women Farmers of California," *San Francisco Bulletin*, 25 December 1904, Harriet Williams Russell Strong Collection, Huntington Library.

58. "A Tour Afoot: Notes of a Trip through Southern California," unidentified newspaper article [*Los Angeles Times?*], 20 June 1891, Harriet Williams Russell Strong Collection, box 18, Huntington Library.

59. Strong left virtually no record of her participation in the organization.

60. *California Outlook*, 22 March 1913, Strong Collection, Huntington Library.

61. Charles Collins Teague, *Fifty Years a Rancher: The Recollections of Half a Century Devoted to the Citrus and Walnut Industries of California and to Furthering the Cooperative Movement in Agriculture* ([Los Angeles]: Charles Collins Teague, 1944), 3, 11, 24, 30, 46.

62. Ibid.; Tomás Almaguer, *Racial Fault Lines: The Historical Origins of White Supremacy in California* (Berkeley and Los Angeles: University of California Press, 1994), 95–97.

63. Ibid., ii.

64. Forrest Crissey, *Where Opportunity Knocks Twice* (1910; reprint, Chicago: Reilly and Britton, 1914), 177–79.

65. Ibid.

66. Ibid.

67. M. B. Waite, "Fruit Growing," *Yearbook of the United States Department of Agriculture, 1904* (Washington, D.C.: Government Printing Office, 1905), 169–72.

68. Ibid., 171.

69. Ibid.

70. Edward F. Adams, "The Position of the Farmer in Our Economic Society," in Edwin G. Nourse, ed., *Agricultural Economics*, 75.

71. Edward F. Adams, *Modern Farmer*, 16.

72. In 1856, the operation grossed two hundred thousand dollars. Ernest Seyd, *California and Its Resources* (London, Trübner and Company, 1859), 125–26.

73. William Brewer, *Up and Down California*, 178, 188, 300.

74. Ibid., 300.

75. LaRue, "Opening Address," *Transactions of the California State Agricultural Society*, 18.

76. Ibid.

77. Ibid., 18–19; for a discussion of specialization as it concerns California agriculture after World War II see Margaret FitzSimmons, "The New Industrial Agriculture: The Regional Integration of Specialty Crop Production," *Economic Geography* 62 (October 1986): 334–53.

78. United States Department of Agriculture, Bureau of Soils, *Reconnaissance Soil Survey of the Lower San Joaquin Valley, California*, (Washington, D.C.: Government Printing Office, 1918), 32; E. J. Wickson, *Rural California* (New York: Macmillan, 1923), 31–32. According to Wickson, settlers found that "in some mysterious way the soil had been 'weathered' all the way down as shown by the fact that earth thrown out in digging ordinary wells or cellars would grow as good plants as the old undisturbed surface." E. W. Hilgard, *Soils: Their Formation, Composition, and Relations to Climate and Plant Growth* (New York: Macmillan, 1906), passim.

79. Hans Jenny and collaborators, "Exploring the Soils of California," in Claude B. Hutchinson, ed., *California Agriculture* (Berkeley and Los Angeles: University of California Press, 1946), 317–93. California's soils seemed so different from those observed in other regions that even as late as 1935 a major study could not associate California's soils with representative types of the great climatic soil groups of the world.

80. R. W. C. Farnsworth, *A Southern California Paradise (in the Suburbs of Los Angeles)* (Pasadena: R. W. C. Farnsworth, 1883), 9

81. Ibid.; *California Horticulturist and Floral Magazine*, 1 (November 1870).

82. "Not only must the kind of fruit to plant be determined by local observation and experience, but often varieties of these fruits must be chosen with reference to adaptation to local environment." Edward J. Wickson, *The California Fruits and How to Grow Them*, 6th ed. (San Francisco: Pacific Rural Press, 1912), 10, 17; the *California Cultivator* wrote, "Where prunes grow famous for their quality, a short walk down the road prune orchards are more valuable as a wood pile." *California Cultivator* 33 (23 September 1909); George C. Roeding, *Roeding's Fruit Growers' Guide* (Fresno: George Roeding, 1919), 3; C. W. Woodworth, "Cooperative Work in Economic Entomology," ([Washington, D.C.]: United States Department of Agriculture, Office of Experiment Stations, 1903), 187.

83. Eugene W. Hilgard, *Address on Progressive Agriculture and Industrial Education* (Jackson, Miss.: Clarion Book, 1873), 22.

84. Ibid.; Fitzgerald, *Business of Breeding*, 81.

85. E. W. Hilgard to the editor of the American Grange Publishing Company, Cincinnati, Ohio, 9 November 1906, letter book "January to November 1906," Records and Papers of the University of California College of Agriculture, Bancroft Library. The first report of the station is dated 1877. Hilgard's first report appeared in January 1884. Verne A. Stadtman, *The University of California, 1868–1968* (New York: McGraw-Hill Book Company, 1970), 145–47.

86. Eugene W. Hilgard, *Report on the Physical and Agricultural Features of the State of California, with a Discussion of the Present and Future Cotton Production in the State* (San Francisco: Pacific Rural Press, 1884), 21, 38, 16; Charles Howard Shinn, "Experimental Agriculture in California," *Sunset* (November 1901): 18.

87. Teague, *Fifty Years a Rancher*, 71.

88. Teague, "The Lemon," an essay delivered before the Farmers Institute, held at Santa Paula, 27 and 28 August 1902, C. C. Teague papers, letters sent, 1901–1905.

89. Hunt, "Suggestions to the Settler," 15, 19.

90. Ibid., 49; R. L. Adams, *Farm Management*, 5.

91. Hunt, "Suggestions to the Settler," 49; Adams, *Farm Management*, 5.

92. United States Department of Agriculture, Weather Bureau, *Climatic Summary of the United States: Section 17—Central California* (Washington, D.C.: Government Printing Office, 1934), 17.

93. Charles C. Colby, "The California Raisin Industry—a Study in Geographic Interpretation," *Annals of the Association of American Geographers* 14 (June 1924): 83; United States Department of Agriculture, Bureau of Soils, *Reconnaissance Soil Survey of the Lower San Joaquin Valley*, 15.

94. One acre-foot is one acre flooded to a depth of one foot.

95. Frank Adams, "Essential Irrigation Factors for a California Farm," typed manuscript, Adams Collection, folder 46, Water Resources Center Archives, Berkeley, California.

96. Winkler, *General Viticulture*, 90–92, 335–36.

97. Wickson, *California Fruits*, 28–32; Hilgard, *Report on the Physical and Agricultural Features;* A. T. Strahorn, United States Department of Agriculture, Bureau of Soils, Field Operations, *Soil Survey of the Fresno Area* (Washington, D.C.: Government Printing Office, 1914), 2089–266. The soils found on the eastern side of the trough of the Central Valley consisted of a granite parent material.

98. Winkler, *General Viticulture*, 54, 564; Heine, interview.

99. Jancis Robinson, *Vines, Grapes, and Wines* (New York: Knopf, 1986), 263, 187. Heine, interview; Winkler, *General Viticulture*, 65. For the way settlers reacted to the diversity of soils see Wickson, *Rural California;* Walter W. Weir and R. Earl Storie, "A Rating of California Soils," *University of California College of Agriculture, Agricultural Experiment Station Bulletin 599* (January 1936); "Ten or fifteen years ago raisin vineyards were planted in many parts of California," stated one grower to a statewide convention, "but experience has shown that the country around Fresno is peculiarly favored in the matter of climate." A. B. Spreckels, "Raisins," *Transactions of the State Agricultural Society during the Year 1899* (Sacramento: State Printing Office, 1900), 13–14.

100. R. L. Adams, *Farm Management Notes*, 24. "Good" results for apples were four hundred packed boxes per acre at 44 pounds per box, or 17,600 pounds per acre, or 8.8 tons.

101. Ibid., 24, 27, 96. Apricots were harvested between June and August, Gravenstein apples between July and August, peaches between July and September.

102. Gustav Eisen, "Why Some Raisin Vineyards Pay More Than Others," *California—A Journal of Rural Industry* 3 (22 March 1890). `

103. N. P. Chipman, *Report upon the Fruit Industry of California* (San Francisco: State Board of Trade of California, 1889), 9.

104. United States Department of Commerce, Bureau of the Census, *Thirteenth Census of the United States, Abstract of the Census* (Washington, D.C.: Government Printing Office, 1913), 24.

105. Ulysis Prentiss Hedrick, *A History of Agriculture in the State of New York* (New York: New York State Historical Society, 1933), 381–93.

106. Ibid. The author does not specify the size of the hothouse, the age of the vines, or any other information. For a history of horticulture in New England among gentleman farmers, see Tamara Plakins Thornton, *Cultivating Gentlemen: The Meaning of Country Life among the Boston Elite, 1785–1860* (New Haven: Yale University Press, 1989).

107. Ulysis Prentiss Hedrick, "The Fruit Districts of New York," *The Fruit Industry of New York, Department of Agriculture of the State of New York Bulletin No. 79* (January 1916): 642–44.

108. Hedrick, "Fruit Districts of New York," 644.

109. Ralph S. Tarr, "Geological History of the Chautauqua Grape Belt," *Cornell University Agricultural Experiment Station Bulletin 109* (January 1896): 121.

110. Ralph S. Tarr, "Decline of Farming in Southern-Central New York," *Bulletin of the American Geographical Society* 41 (1909): 275; E. G. Lodeman, "Some Grape Troubles of Western New York," *Cornell University Agricultural Experiment Station Bulletin 76* (November 1894).

111. N. P. Chipman, "Wheat vs. Fruit," *Proceedings of the Tenth State Fruit Growers' Convention . . . November 23, 1888* (Sacramento: Superintendent of State Printing, 1889), 102–7.

112. Milton Thomas, "Profits of Fruit Raising," *Proceedings of the Tenth State Fruit Growers' Convention . . . November 23, 1888* (Sacramento: Superintendent of State Printing, 1889), 85–93.

113. The speaker must have known that orange growers commonly used smudge pots, oil-burning stoves, to raise the temperature and save their crops on chilly nights.

114. N. P. Chipman, *Annual Report of the Committee on the Industrial Resources of the State, Annual Report of the California State Board of Trade* [Sacramento, 1892], 3.

115. Ibid., 16–17; Thomas J. Vivian, *The Commercial, Industrial, Agricultural, Transportation, and other Interests of California* (Washington: Bureau of Statistics, Treasury Department, 1891), 326, 255–86; *Pacific Rural Press* 101 (1 January 1921); M. Theodore Kearney, "Fresno County, California: The Land of Sunshine, Fruits, and Flowers," in *How to Make Money in California* (Fresno: self-published, 1893).

116. *Fresno County, California—Where Can Be Found Climate, Soil, and Wa-*

ter, the Only Sure Combination for the Vineyardist ([San Francisco]: Vogel, Hall, Lisenby), 1887.

117. Colby, "California Raisin Industry," 50. Fresno County's grape harvest accounted for 1,817,000 tons in 75,053 freight cars. "Produce and Movement of Fruits and Vegetables in the United States," *Bureau of Railway Economics Bulletin No. 20* (April 1927): 6.

118. Chipman, *Annual Report*, 3, 9.

119. Danhof, *Change in Agriculture*, 9–26.

120. Sucheng Chan, *This Bitter-Sweet Soil: The Chinese in California Agriculture, 1860–1910* (Berkeley and Los Angeles: University of California Press, 1986), 79, 90.

121. Seyd, *California and Its Resources*, 125–26; growers shipped 713 tons of grapes to San Francisco in 1857 and 675 tons in 1859. Thompson and West, *History of Los Angeles County, California* (Oakland: n.p., 1880), 94, 96, quoted in Paul W. Gates, ed., *California Ranchos and Farms, 1846–1862, Including the Letters of John Quincy Adams Warren* (Madison: State Historical Society of Wisconsin, 1967), 73n.

122. California Fruit Growers, *Official Report of the Fourth Annual State Convention of California Fruit Growers at San Francisco . . . September 1884* (San Francisco: Pacific Rural Press, 1884), 2.

123. *California Farmer*, 32 (1869), quoted in Erich Kraemer and H. E. Erdman, "History of Cooperation in the Marketing of California Fresh Deciduous Fruits," *University of California College of Agriculture Agricultural Experiment Station Bulletin 557* (September 1933): 7n.

124. *Pacific Rural Press* 45 (4 and 12 March 1893).

125. Ibid.

126. *Pacific Rural Press* 50 (19 February 1898).

127. George Kellogg, *Private Telegraph Code of George Kellogg, Grower and Shipper of Choice Mountain Fruit, New Castle, Placer County, California* (Sacramento: H. S. Crocker, 1893), 14–15.

128. Ibid., 23–24.

129. *New York Sun*, 22 September 1889, quoted in Chipman, *Report upon the Fruit Industry*, 24–25.

130. *Pacific Rural Press* 32 (23 October 1886).

131. Ibid., 24. The difference in price was astonishing but so was the difference in volume. California fruits sold high, but not in great quantities. See the following chapter.

132. For the following description I have intentionally taken a young vineyard and conservative statistics to demonstrate the profitability of raisin growing. Although vines are fairly predictable, it is very difficult to assess the profitability of a vineyard. Local conditions, yearly markets, and unexpected misfortune figure in any agricultural enterprise.

133. These prices were compiled in 1890 by Gustav Eisen, an early expert and observer of the raisin industry. Eisen, *Raisin Industry*, 170–74; also see "Fresno County, California," 37, 71. Prices per box in the *Pacific Rural Press* fluctuated between $3.50 in June of 1871 to a range of $1.30 to $2.00 in August of 1892. *Pacific Rural Press* 1 (3 June 1871) and 44 (27 August 1892).

134. *Pacific Rural Press* 1 (3 June 1871) and 44 (27 August 1892).

135. Promoters often combined all labor and processing costs into one figure: picking, packing, and "putting upon on the market" was assessed at $40 to $50 per ton in the late 1880s and early 1890s. *Fresno County, California*, 71, 65; Baker, *76 Land and Water Company*, 24–25. *Imperial Fresno*, 20. For the wages paid to laborers, see "Horticulture in California," in United States Department of Commerce, Bureau of the Census, *Eleventh Census of the United States—Agricultural*, passim.

136. California Fruit Growers, *Report . . . 1884*, 9; Thomas Forsyth Hunt estimated that farmers and fruit growers in California could expect an income equal to 25 percent of their invested capital. Thomas Forsyth Hunt, "Suggestions to the Settler," 4–5.

137. The statistics are from 1909. Ray Palmer Teele, *Irrigation in the United States* (New York: Appleton and Company, 1915), 48–49.

138. Gustav Eisen, "Raisin Industry of Fresno"; Thompson, *Official Atlas Map of Fresno County*, 1, 13; *Pacific Rural Press* 1 (14 January 1871).

139. Chan, *This Bitter-Sweet Soil*, 322; "Fresno County, California," 65.

140. *Pacific Rural Press*, 45 (21 January 1893).

141. Ibid., 45 (14 January 1893) and 45 (1 April 1893).

142. Alfred D. Chandler Jr., with Takashi Hikino, *Scale and Scope: The Dynamics of Industrial Capitalism* (Cambridge: Belknap Press, Harvard University, 1990), 17. I will return to this subject a number of times in the following chapters.

143. J. Karl Lee, "Economies of Scale of Farming in the Southern San Joaquin Valley, California," (manuscript produced by the United States Department of Agriculture Bureau of Agricultural Economics, Berkeley, April 1946), 94; R. L. Adams, "The Management of Large Farms," *Agricultural Engineering* 12 (September 1931): 354–57.

144. Ibid., 54. For more about large-scale production see chapter 6. For details on the advantages of intensive production see [Harry Drobish?], *Summary of Enterprise Efficiency Study—Peaches, for Tulare County—1927*, manuscript report for the Agricultural Extension Service, Drobish Papers, Bancroft Library, 4–11. J. Karl Lee's study of specialized fruit orchards during 1936–40 determined that farms under 100 acres had an index number of 160 measuring their intensity of production while farms of between 100 and 479 acres indexed at 149. The study did not distinguish between different kinds of fruit.

145. Hogue, Murray, and Sesnon, *Fresno County, California* (1889), 17. M. B. Waite of the United States Department of Agriculture wrote, "There is no doubt a great advantage [regarding large scale operations] in marketing fruit, especially peaches and apples, in carload lots." Waite, "Fruit Growing," 170–71.

146. *Official Map of the County of Yolo, California* (San Francisco: P. N. Ashley, 1909). Property listed under the name of Mary J. Blowers.

147. "Ledger," dated June 1900, R. B. Blowers and Sons Papers, Bancroft Library. The names of other growers who shipped through Blowers are listed in "Prices Realized: Fruit Sold for Account of Sgobel and Day at Erie R.R., Pier 20 N.R. Sold By Brown and Seccomb, Auctioneers," (New York, 8 October 1908), R. B. Blowers Papers, Carton 5, Bancroft Library, University of California, Berkeley. Not all of the names in the price sheets are found on the county map printed in 1909. One reason might be that owners of colony plots as small as ten

and twenty acres were not listed individually. See *Official Map of the County of Yolo, California* and *Official Map of the County of Solano, California* ([San Francisco?]: E. N. Eager, 1909).

148. Letter, Macheca Bros. to A. B. Blowers, 8 November 1887, R. B. Blowers Papers, Bancroft Library; Letter, George Scott, New England Agent of the Placer County Mountain Fruit Company, to R. B. Blowers, 18 October 1913.

149. Robert DiGiorgio and Joseph A. DiGiorgio, *The DiGiorgios: From Fruit Merchants to Corporate Innovators*, an oral history conducted in 1983 by Ruth Teiser, Regional Oral History Office, the Bancroft Library, University of California, Berkeley, 1986, 6.

150. *New York Times*, 26 February 1951, pp. 24:2.

151. *Fresno County, California, Its Offering for Settlement; Imperial Fresno*, 15.

152. A. B. Butler, "California and Spain;" Rohrabacher, *Fresno County*, 47–48.

153. Gustav Eisen, "Raisin Industry of Fresno." These packinghouses hired between 250 and 400 workers each season.

154. Clarence E. Edwards, *Transactions of the State Agricultural Society for the Year 1908* (Sacramento: State Printing Office, 1909), 29.

155. Carl S. Scofield, "The Present Outlook for Irrigation Farming," *Yearbook of the Department of Agriculture, 1911* (Washington, D.C.: Government Printing Office, 1911), 374.

156. George C. Roeding, "Plea for Diversified Agriculture," *Transactions of the California State Agricultural Society for 1909* (Sacramento: State Printing Office, 1910), 22.

3.   Organize and Advertise

1. The state's fruit crop was valued at $48,718,000 in 1909. In the same year, the combined value of all field crops, including onions, sweet potatoes, and sugar beets, came to $95,757,000. From *Economic Resources and Extractive Industries of California*, quoted in Robert Glass Cleland and Osgood Hardy, *March of Industry* (Los Angeles: Powell Publishing, 1929), 264.

2. Harris Weinstock, "The Eastern Fruit Market," *Pacific Rural Press* 32 (25 September 1886).

3. *Pacific Rural Press* 30 (3 October, 21 October 1885); Erich Kraemer and H. E. Erdman, "History of Cooperation," 16. The question of whether or not shippers should be allowed to join the union divided its membership and eventually destroyed it. Adams, *Modern Farmer*, 452–53. The Fruit Union was established as a stock corporation with an authorized capital stock of $250,000, but actual subscriptions for stock (at $1.00 per acre of orchard land) never exceeded $15,000.

4. Cronon, *Nature's Metropolis*, 120–32.

5. Teague, *Fifty Years a Rancher*, 52–53.

6. [California State Horticultural Society], *Tabulated Statement Showing Shipments of California Green Deciduous Fruit Made to Eastern Markets, and the Results of Its Sale Therein* ([Sacramento]: 1902). The statement gives record of all sales for an eight-day period in July 1902, with the profits and losses on each carload. It also includes records for 3,000 cars shipped from 5 July to 30 August

1902. Carloads of fruit shipped from cities and towns throughout northern California (mostly from the Sacramento Valley) required an average of 10 to 11 days to reach Chicago and 14.5 days to reach New York. Also see California Fruit Growers and Shippers Association, Transportation Committee, *Memorial from the Fruit Growers of California to the Transportation Companies, Asking for a Quicker Time to Points in the East, and for Amendment of the Minimum Carload* ([Sacramento?], 1901), 6

7. Cronon, *Nature's Metropolis*, 89–90.

8. Ibid.

9. Fred Wilbur Powell, "Co-operative Marketing of California Fresh Fruits," *Quarterly Journal of Economics* 25 (February 1910): 406.

10. *Memorial from the Fruit Growers*, 7–8. The cars had to be re-iced at regular points along the way: Sparks, Carlin, Ogden, etc. California Growers and Shippers Protective League, *Refrigeration Test Trip of Six Cars of Plums and Other Deciduous Fruits from Central and Northern California Points to Chicago and New York City, June 9th to July 1, 1924* (San Francisco: California Growers and Shippers Protective League, 1924).

11. Neither Sgobel and Day nor R. B. Blowers considered the refrigerated car a foolproof way of shipping across the country. Sgobel and Day wrote to their client in California, "You would be surprised to see how quickly this fruit decays after it comes out of a refrigerated car. . . . " Sgobel and Day to R. B. Blowers and Sons, 24 October 1896, R. B. Blowers Papers, Bancroft Library.

12. Weinstock, "The Eastern Market for California Green Fruits," *Pacific Rural Press* 32 (23 October 1886).

13. California Fruit Growers and Shippers Association, *Memorial from the Fruit Growers*, 7.

14. Ibid.

15. Ibid., 8. Emphasis in original; Henry E. Erdman, *American Produce Markets* (Boston: D. C. Heath and Company, 1928), 296–97.

16. Harris Weinstock, "The Eastern Market for California Green Fruit," 9 October 1886 and 23 October 1886.

17. Ibid.; in 1929, 85 percent of all the fruits and vegetables that fed the New York metropolitan district passed through the New York Terminal Market. M. P. Rasmussen, "Some Facts Concerning the Distribution of Fruits and Vegetables by Wholesalers and Jobbers in Large Terminal Markets," *Cornell University Agricultural Experiment Station Bulletin 494* (October 1929): 18–23.

18. The cast of characters in the fruit business was identical to that in other industries where manufacturers did not market their own products. Here is a glossary of terms.

> Shipper: Anyone who attempted to sell fruit to a distant market. A shipper could be an individual grower like George Kellogg or a larger organization like R. B. Blowers.
>
> Commission Merchants: Selling agents who consigned fruit from shippers and sold to wholesale or retail establishments. They paid the shipper or grower only after the sale and took a percentage of the profits, usually about 7 percent; also called a consignment merchant or a broker.
>
> Jobber or Buyers: Wholesale fruit dealers who bought from producers, brokers, or commission merchants and sold to the retail trade.

Buyer: Anyone who purchased fruit wholesale and sold it to the public. Grocery stores, restaurants, or jobbers might be buyers. Each of these performed the function of every other at different times.

See Glenn Porter and Harold C. Livesay, *Merchants and Manufacturers: Studies in the Changing Structure of Nineteenth-Century Marketing* (Baltimore: Johns Hopkins University Press, 1971), passim; G. Harold Powell, *Cooperation in Agriculture* (New York: Macmillan, 1913; reprint, 1919), 200–203; William W. Cumberland, *Cooperative Marketing: Its Advantages as Exemplified in the California Fruit Growers Exchange* (Princeton, N.J.: Princeton University Press, 1917), 51.

19. Powell, "Co-operative Marketing," 397; Porter and Livesay, *Merchants and Manufacturers*.

20. Understood another way, in newly settled irrigated regions with an expanding economy farmers could virtually charge the cost of production *plus* any freight costs. These were the fat years. Then the surging number of acres under cultivation forced producers into long-distance trade where merchants had discretion and control. In these circumstances the price equaled what buyers would pay for it *less* the cost of freight and commission charges—a complete reversal of early conditions. Carl S. Scofield, "The Present Outlook for Irrigated Farming," United States Department of Agriculture, *Yearbook* (Washington D.C.: Government Printing Office, 1911), 371–82.

21. For an excellent example of this relationship as it concerned a major New York commission house, Sgobel and Day, and a shipper of deciduous fruits in the Sacramento Valley, R. B. Blowers and Sons, see the Blowers Papers, Bancroft Library.

22. M. Theodore Kearney, "Address among the Farmers as Applied to the Raisin Industry," *Official Report of the Twenty-Fourth State Fruit-Growers' Convention* (Sacramento, 1900), 60. "Thereafter," wrote Kearney, "the dealers in the East refused to pay for raisins except after arrival and upon examination; and . . . the growers were obliged to ship the raisins East on consignment." William W. Cumberland, *Cooperative Marketing*, 51; Teague, *Fifty Years a Rancher*, 74; John William Lloyd, *Cooperative and Other Methods of Marketing California Horticultural Products*, University of Illinois Studies in the Social Sciences, 8 (March 1919).

23. *Pacific Rural Press*, 45 (15 April 1893).

24. *California Fruit Grower*, 25 December 1890; *California Fruit Grower*, January 1, 1891, in Gustav Eisen Papers, Manuscript Collection, Bancroft Library.

25. *California Fruit Grower*, 25 December 1890, in Gustav Eisen Papers, Manuscript Collection, Bancroft Library.

26. Ibid.; M. Theodore Kearney, "Address among the Farmers," 60; Edward F. Adams, a farmer and businessman who worked for the University of California with its Farmers' Institutes, wrote that "[t]he duty of the commission merchant is to faithfully prosecute the business of his consignor to the best of his ability . . . , making immediate remittance of the net proceeds." Farmers were advised to choose their agents with care and to avoid borrowing money from them. Adams, *Modern Farmer*, 152–7.

27. Cumberland, *Cooperative Marketing*, 45.

28. H. [Herbert] A. Fairbank to R. B. Blowers and Sons, 4 September 1892 and 24 September 1892, R. B. Blowers Papers, Bancroft Library; Teague, *Fifty Years a Rancher*, 75–76.

29. Weinstock, *Pacific Rural Press*, 32 (23 October 1886).

30. In 1896 the Erie Railroad built an auction room with a capacity large enough to sell all the fruit passing through New York at any given time. *Sgobel and Day Fruit Distributors* (New York: Sgobel and Day, 1922); Weinstock telegraphed R. B. Blowers in Woodland about the outcome of the first auction sale: "In spite of many unfavorable conditions first auction sale did well . . . look for permanent success[—]excellent." Telegraph message, Harris Weinstock in Chicago, Illinois to R. B. Blowers, 21 June 1888, all sources in Blowers Papers, Bancroft Library.

31. Weinstock, "Eastern Market for California Green Fruits," 23 October 1886.

32. Weinstock, "When to Ship Fruit," *Pacific Rural Press* 32 (2 October 1886). Years later, Carl Scofield noted that the newly irrigated lands of the West "will meet keener competition than previously from eastern lands." Carl S. Scofield, "Present Outlook," 375–76.

33. Weinstock, "When to Ship Fruit."

34. Porter and Livesay, *Merchants and Manufacturers*, passim. See especially chapter 8, "The Manufacturer Ascendant: Changing Markets in Producers' Goods." Two factors resulted in the end of the commission house as an independent business enterprise and the development of producer-owned selling networks: the special demands of new consumer products and the concentration of markets into easily identifiable urban areas. Both conditions existed in the fruit industry by World War I.

35. [California State Horticultural Society]. *Tabulated Statement Showing Shipments*, passim.

36. Kearney, "Address among the Farmers," 60; A. B. Spreckels, "Raisins," *Transactions of the State Agricultural Society during the Year 1899* (Sacramento: Office of State Printing, 1900), 13–14.

37. California Fruit Growers and Shippers Association, *Memorial from the Fruit Grower*, 7. Not my emphasis.

38. Kearney, "Address among the Farmers," 60.

39. T. S. Van Dyke quoted in William Smythe, *Conquest of Arid America*, 128.

40. *Pacific Rural Press* 55 (1 January 1898).

41. Ibid.

42. Martin Ridge, *Ignatius Donnelly;* Lawrence Goodwyn, *Populist Moment*, 168.

43. Theodore Roosevelt, "Seventh Annual Message, December 3, 1907," vol. 3 (1905–66) of *The State of the Union Messages of the Presidents, 1790–1966*, edited by Fred L. Israel (New York: Chelsea House Publishers, 1967), 2263.

44. Butterfield, *Chapters in Rural Progress*, 14.

45. Edwin G. Nourse, "The Economic Philosophy of Co-operation," *The American Economic Review* 12 (December 1922): 597. Also by Nourse, *The Legal Status of Agricultural Cooperation* (New York: Macmillan, 1927); *Agricultural Economics;* and *Marketing Agreements under the Agricultural Adjustment Act* (Washington, D.C.: Brookings Institution, 1935).

46. By 1927 Americans assumed that farm protest had become part of the nation's past. Charles A. and Mary R. Beard, for example, noted, "Agriculture had passed out of the age of mere uproarious protest into the age of collective effort

and constructive measures." Charles A. Beard and Mary R. Beard, "The Industrial Age," vol. 2 of *The Rise of American Civilization* (New York: Macmillan, 1927), 284.

47. Mary Neth, *Preserving the Family Farm: Women, Community, and the Foundations of Agribusiness in the Midwest, 1900–1940* (Baltimore: Johns Hopkins University Press, 1995), 102–7.

48. Bailey, *The Country-Life Movement*, 127.

49. Ibid., 106, 125.

50. Victoria Alice Saker, "Benevolent Monopoly: The Legal Transformation of Agricultural Cooperation, 1890–1943" (Ph.D. diss., University of California at Berkeley, 1990), 18–22; Herman Steen, *Coöperative Marketing: The Golden Rule in Agriculture* (Garden City, N.Y.: Doubleday, Page, and Company, 1923), 3.

51. G. Harold Powell, *Coöperation in Agriculture*, 27, 40–44; Saker, "Benevolent Monopoly," 18–22; Steen, *Coöperative Marketing*, 8–10.

52. G. Harold Powell, "Fundamental Principles of Co-operation in Agriculture" *University of California College of Agriculture, Agricultural Experiment Station Circular No. 123* (October 1914): 2.

53. The California Associated Raisin Company, or Sun-Maid, did not fit the model of the cooperative as outlined by advocates of cooperation. Sun-Maid held capital stock, and nonmembers could subscribe. Saker, "Benevolent Monopoly," 224.

54. Powell died at the age of fifty, in 1922. For some of Powell's observations on the orange business see Powell, *Letters from the Orange Empire*, edited by Richard G. Lillard (Los Angeles: Historical Society of Southern California, 1990), 1–7; Powell, "The Handling of Fruit for Transportation," United States Department of Agriculture *Yearbook* (Washington, D.C.: Government Printing Office, 1905). Also see H. Vincent Moses, "'The Orange-Grower Is Not a Farmer': G. Harold Powell, Riverside Orchardists, and the Coming of Industrial Agriculture, 1893–1930," *California History* 74 (spring 1995): 22–37.

55. In the conception of Charles W. Macune, the subtreasury provided growers with greenback advances, or government-issued "certificates" that could be traded as legal notes. The subtreasury eliminated the need for credit banks, furnishing merchants, and mortgage companies in the business of farming. The idea was to put the government into agriculture on the side of the farmer and to stockpile farm commodities in the national purse. See Goodwyn, *Democratic Promise;* Ridge, *Ignatius Donnelly;* John D. Hicks, *The Populist Revolt: A History of the Farmers' Alliance and the People's Party* (Minneapolis: University of Minnesota Press, 1931); Fred A. Shannon, *Farmer's Last Frontier*. For an account of public anxiety over the perceived closing of the frontier, refer to Wrobel, *End of American Exceptionalism*. On the depression of the 1870s see Herbert G. Gutman, "Social and Economic Structure and Depression in American Labor in 1873 and 1874" (Ph.D. diss., University of Wisconsin, 1959).

56. Powell, *Cooperation in Agriculture*, 22.

57. Ibid., 1.

58. Ibid., 2.

59. Ibid., 5.

60. Ira B. Cross, "Cooperation in Agriculture," *American Economic Review* 1 (September 1911): 543–44.

61. The act of 1915 stipulated that the market director conduct a *commission business* for all farmers and ranchers. The director's job was to act as head commission merchant for every bottle of milk, carton of eggs, and sack of flour west of the Sierra. Weinstock thought the whole idea absurd. He asked for Johnson's help to persuade the legislature, and by an act of 1917 the State Commission Market became the State Market Commission. Harris Weinstock, *Statement of Harris Weinstock, State Marketing Director in Answer to the Criticisms of Senator William E. Brown of Los Angeles Relative to the Administration of the California State Marketing Law* (San Francisco: self-published, 1917); *First Annual Report of the State Market Director of California to the Governor of California, for the Year Ending December 1, 1916* (Sacramento: State Printing Office, 1916), 7–8; "Has John P. Irish an Onion in His Eye?" *Pacific Rural Press* (2 February 1916); Harris Weinstock to Governor Hiram Johnson, 18 January 1917, in Chester Rowell Papers, Letters to Rowell, Bancroft; Weinstock to Johnson, 15 February 1916, Hiram W. Johnson Papers, Bancroft Library; *Second Annual Report of the State Market Director*, 5–10, Appendix B, "State Market Commission Act of California [Approved 1 June 1917]"; "History of the Bureau of Markets," in the Records of the Department of Agriculture, "Dr. Braun's Report File," California State Archives; Weinstock to David Lubin, 20 August 1917, David Lubin Papers, 1848–1919, Western Jewish History Center, Juda L. Magnes Museum, Berkeley, California.

62. *Second Annual Report of the State Market Director*, 15.

63. Note cards, H. E. Erdman Papers, Box 27, Deciduous Fruits by County, "Contra Costa."

64. Rahno Mabel MacCurdy, *The History of the California Fruit Growers Exchange* (Los Angeles: self-published, 1927), 26–27.

65. Dean Witter and Company, *California Packing Corporation: A Study of Impressive Progress* (San Francisco: Dean Witter and Company, 1950), 14; Vicki Ruiz, *Cannery Women; Cannery Lives: Mexican Women, Unionization, and the California Food Processing Industry, 1930–1950* (Albuquerque: University of New Mexico Press, 1987).

66. "Calpak: The Adventures of Del Monte Brand," *Fortune* (November 1938); "California Packing Corp. Operates on a Vast Scale," *San Francisco Chronicle* 28 February 1930; "Annual Report of the California Packing Corporation," *Western Canner and Packer* 10 (May 1918); Dean Witter and Company, *California Packing Corporation*, 14.

67. California Pear Growers' Association, Minutes, 2 April, 29 April, 6 May, 15 May, 24 May, 10 June, 17 June 1918, Bancroft Library.

68. California Pear Growers' Association, Minutes, 17 June 1918, Bancroft Library.

69. Frank Swett to H. K. Donaldson, 10 April 1923, manuscript letter in H. E. Erdman Papers, carton 16, folder: "Pears/Frank Swett," Bancroft Library.

70. "Summary of an Interview with Frank T. Swett;" California Pear Growers' Association, Minutes, 5 May 1920.

71. "Report of the Board of Directors to the Members of California Pear Growers' Association," *Third Annual Report and Statement, Season 1920* (San Francisco: California Pear Growers' Association, 1920), 4.

72. "Summary of an Interview with Frank T. Swett Conducted by H. E. Erd-

man and a Group of Graduate Students," March 1930, H. E. Erdman Papers, carton 16, folder "Frank Swett," Bancroft; *1921 Roster of Members of the California Pear Growers' Association* (San Francisco: California Pear Growers' Association, 1921).

73. New York, Washington, and Oregon grew 4,372,000 bushels, and California alone produced 4,600,000 bushels, or 115,000 tons. At $72 a ton, the United States Department of Agriculture valued the California harvest at $8,280,000. *Agriculture Yearbook, 1925* (Washington, D.C.: Government Printing Office), 876, 884; Steen, *Coöperative Marketing*, 69–70.

74. *Western Canner and Packer* 12 (November 1920).

75. R. E. Hodges, "Canneries to Balance Fruit Markets," *Pacific Rural Press* 99 (10 January 1920).

76. California Pear Growers' Association, Minutes, 17 January 1921.

77. *Western Canner and Packer* 12 (August 1920).

78. Ibid.

79. Erdman, "Summary of an Interview with Frank T. Swett."

80. *Western Canner and Packer* 12 (November 1920).

81. Frank T. Swett, "California Agricultural Cooperatives, Interview with Frank T. Swett." Conducted by Willa Klug Baum (Berkeley: Regional Oral History Project, 1968); Henry E. Erdman, *American Produce Markets*, 82–85.

82. In 1924 the Southern Pacific sponsored a test trip of deciduous fruits from California to New York to observe how produce of all kinds survived the transcontinental journey. The results helped the railroad to provide better service and gave accurate information about the dependability of refrigeration. California Growers and Shippers Protective League, *Refrigeration Test Trip*.

83. R. G. Phillips, *Wholesale Distribution of Fresh Fruits and Vegetables* (Rochester, N.Y.: Joint Council of the National League of Commission Merchants of the United States, the Western Fruit Jobbers' Association of America, and the International Apple Shippers' Association, 1922), 3.

84. Sgobel and Day to R. B. Blowers and Sons, 19 November 1915, R. B. Blowers and Sons Papers, Bancroft Library.

85. California Pear Growers' Association, *More Buyers for More Pears*, ([Sacramento]: California Pear Growers' Association, November 1923), 2–3. This pamphlet appears to have been intended to persuade growers inside the association to continue their support of the advertising campaign and to attract new members.

86. Sgobel and Day to Miss Carrie Blowers, 14 September 1892 and Sgobel and Day to R. B. Blowers and Sons, 1 October 1894, Blowers Papers, Bancroft Library. Also see the *Pacific Rural Press* for 1893.

87. California Pear Growers' Association, Minutes, 17 January 1921; and California Pear Growers' Association, *Third Annual Report and Statement, Season, 1920*, 4–6.

88. W. S. Killingsworth, "Factors in California's Fruit Expansion," *Pacific Rural Press* 99 (31 January 1920).

89. California Pear Growers' Association, *Third Annual Report and Statement, Season, 1920*, 6.

90. Gordon T. McClelland and Jay T. Last, *California Orange Box Labels: An Illustrated History* (Beverly Hills, Calif.: Hillcrest Press, 1985), 6 and passim.

Labels and brands represented individual growers or groups in local packing associations. Even when selling through a cooperative, these growers and groups still maintained their local brand name.

91. Ibid., 36–40.

92. Ibid., 38.

93. Ibid., 38–45.

94. California Pear Growers' Association, Minutes, 17 January 1921.

95. *The California Pear Grower* 1 (June 1921).

96. "Powell Speaks at Riverside," *The California Citrograph* 2 (November 1916).

97. Ibid.; California Pear Growers' Association, Minutes, 17 January 1921. "Year by year these advertised fruits and nuts have been in constantly increasing demand by the American people and at remunerative prices to the growers. This year the price may not be as good as growers would like, BUT IF IT WERE NOT FOR THE GREATER DEMAND WHICH ADVERTISING HAS BUILT, THERE MIGHT BE NO PRICE AT ALL."

98. *The California Pear Grower* 1 (June 1921).

99. Ibid., (July 1921).

100. California Pear Growers' Association, Minutes, 17 January 1921.

101. Ibid., 22 May 1922.

102. *Packer*, September 1915, quoted in Nourse, ed., *Agricultural Economics*, 114–15.

103. Ibid., 112–13.

104. California Pear Growers' Association, *More Buyers for More Pears*, 2–3.

105. J. Walter Thompson Company, "Presentation on California Fresh Bartlett Pears," Library of the Giannini Foundation for Agricultural Economics, University of California, Berkeley, California, 40.

106. Ibid., 49–59.

107. Ibid., 136–37.

108. Don Francisco, "The Advertising of Agricultural Specialties," *American Cooperation: A Collection of Papers and Discussions Comprising the Fourth Summer Session of the American Institute of Cooperation at the University of California*, vol. 2 (Washington D.C.: American Institute of Cooperation, 1928), 136.

109. J. Walter Thompson Company, "Presentation on California Fresh Bartlett Pears," 2, 138.

110. Ibid., 139.

111. Steen, *Coöperative Marketing*, 9.

112. *More Buyers for More Pears*, 2.

113. Weinstock, "The Eastern Fruit Market."

114. W. L. Crowe, "The Standardization of the Orange in California," *The California Citrograph* 1 (December 1915); G. H. Hecke, "Standardization in Packing of Fruits and Vegetables," *California Fruit News* 61 (7 February 1920).

115. *Pacific Rural Press* 45 (8 April 1893); J. W. Jeffrey, "Standardize the Fruit Pack," in *Proceedings of the Seventh Annual Convention of the Western Fruit Jobbers Association* . . . February 1911 (Denver: Western Fruit Jobbers Association, 1911), 119; Cumberland, *Cooperative Marketing*, 52–53; Erdman, *American Produce Markets*, 246.

116. J. W. Jeffrey, "Standardize the Fruit Pack," 126.

117. B. F. Walton, "Grading and Packing Fruit to Meet the Wants of the Trade," *Pacific Rural Press* 49 (8 June 1895). My emphasis.

118. Ibid.

119. Frank Swett, "Your Eastern Pear Merchandising and Advertising," *California Pear Grower* 4 (October 1924); *California Pear Grower* 2 (July 1922).

120. Erdman, *American Produce Markets*, 251–55.

121. G. H. Hecke, "Standardization in Packing of Fruits and Vegetables," 4.

122. Warren P. Tufts, "The Packing of Apples in California," *University of California, College of Agriculture, Agricultural Experiment Station Circular No. 178* (October 1917): 11; W. S. Killingsworth. "Standardization Promotes Fruit Industry," *Pacific Rural Press* 99 (14 February 1920).

123. California Pear Growers' Association, Minutes, 22 July 1918; Erdman, *American Produce Markets*, 253–57.

124. Erdman, *American Produce Markets*, 266; D. F. Houston, "Report of the Secretary of Agriculture," United States Department of Agriculture *Yearbook* (Washington, D.C.: Government Printing Office, 1917), 29.

125. J. W. Jeffrey, "Standardize the Fruit Pack," 126; letter, C. C. Teague to Waldo A. Hardison, 12 July 1902, C. C. Teague Papers, Bancroft Library.

126. Jeffrey, "Standardize the Fruit Pack," 126.

127. George K. Holmes, "Consumers' Fancies," United States Department of Agriculture *Yearbook* (Washington, D.C.: Government Printing Office, 1905), 417; B. F. Walton, "Grading and Packing Fruit to Meet the Wants of the Trade," *Pacific Rural Press* 49 (8 June 1895).

128. Holmes, "Consumers' Fancies," 420.

129. Ibid., 434.

130. Ibid., 420.

131. George P. Weldon, "Pear Growing in California: A Practical Treatise Designed to Cover Some of the Important Phases of Pear Culture within the State," *Monthly Bulletin of the California State Commission of Horticulture* 7 (May 1918): 21.

132. Ibid., 22–23. Like almost every food grown commercially in California, the Bartlett was an immigrant. It traveled from England, where it was called "William's Bon Chretien." In his *Fruit Trees of America*, Andrew Jackson Downing says that it was propagated in America by Enoch Bartlett of Massachusetts and named after him.

133. Ibid., 61.

134. Ibid., 65. The test regions: northern coast, central coast, southern coast, northern and central interior valley, southern desert, Imperial Valley, Sierra Nevada foothills and mountains, and Modoc and Inyo Counties. Five of the top fourteen flourished in more than one region at the "best" level of production, while the Seckel and the Bosc grew "best" in five and three regions respectively.

135. A. W. McKay et al., "Marketing Fruits and Vegetables, " United States Department of Agriculture *Yearbook* (Washington, D.C.: United States Department of Agriculture, Government Printing Office, 1926), 628.

136. *California Pear Grower* 1 (January 1922); *Pacific Rural Press* 99 (10 January 1920); Adams, *Farm Management*, 100.

4.   A Chemical Shield

1. I have chosen to focus only on insecticides because the urgency of their development made them a constant topic at grower conventions and in experiment station bulletins. For another point of view on the importance of insecticides in the fruit business see Howard Seftel, "Government Regulation," 369–402. Seftel emphasizes the role of state government while the present chapter treats the university and private industry as the central institutions in the rise of chemical pest control. Portions of this chapter are reprinted from Steven Stoll, "Insects and Institutions: University Science and the Fruit Business in California," *Agricultural History* 69 (spring 1995): 216–39, by permission.

2. The word *tule* derives from the Aztec *tullin* or *tollin*, meaning marsh grass. Spaniards used the word to describe any of a number of marsh plants like cattail, bulrush, and the common tule, *Scirpus acutus*. William L. Preston, *Vanishing Landscapes: Life and Land in the Tulare Lake Basin* (Berkeley and Los Angeles: University of California Press, 1981), 22n.

3. John Muir, *Rambles of a Botanist among the Plants and Climates of California* (Los Angeles: Dawson's Book Shop, 1974), 22; William H. Brewer, *Up and Down California*, 202–3, 382.

4. European annual grasses spread rapidly without cultivation and soon replaced perennial bunchgrasses all over California. One key reason for this succession: grazing. Many perennials spread through vegetative reproduction, extending along rhizomes. Although perennials also produce seeds, they tend not to depend on seeds for reproduction. Cattle and sheep consumed perennials faster than the plants could reproduce vegetatively or sexually. Annual grasses, on the other hand, go to seed every season. When perennials declined in numbers, annuals continued to increase—even with grazing—and soon became the most common species in the state. American land speculators who came to the San Joaquin Valley and imagined miles of wheat in place of the waving grasses probably saw European transplants and not the "virgin" flora they supposed. Beecher Crampton, *Grasses in California*, California Natural History Guides, no. 33 (Berkeley and Los Angeles: University of California Press, 1974), 26–33; William S. Chapman, letter of 27 August 1868, in the San Francisco *Bulletin*, published in Gerald D. Nash, "Henry George Reexamined," 133–37.

5. James G. Needham, Stuart W. Frost, and Beatrice H. Tothill, *Leaf-Mining Insects* (Baltimore: Williams and Wilkins, 1928), passim. This work, rather old and out of date, nonetheless provides an excellent explanation of insects and host plants. See especially the list of 422 species and the insects that feed on them.

6. "It is a matter of no little biological interest that the most highly specialized and most numerous leaf-miners do little harm to the plants whose leaves they inhabit. . . . These [insects] have found a way of living that lets the leaf live also." This describes insects long associated with a single plant species. Less specialized bugs eat from many species and do more damage to the plants. In a healthy ecosystem, both live in numbers regulated by predators. Ibid., 36.

7. For a very perceptive discussion of weeds and pests and their relationship to human culture and cultivation see Michael Pollan, *Second Nature: A Gardener's Education* (New York: Bantam Doubleday, 1991).

8. Even plants that can defend themselves against insects in the wild often lose

their immunity when cultivated. Orchard trees receive more water and develop softer, more succulent leaves than wild plants. Ian Sussex, professor of biology and plant genetics, University of California at Berkeley, interview by author, 3 March 1993.

9. Charles S. Elton, *The Ecology of Invasions by Animals and Plants* (New York: John Wiley and Sons, 1958; reprint, 1977), 15 and passim; Mary Louise Flint and Robert van den Bosch, *Introduction to Integrated Pest Management* (New York: Plenum, 1981), 139–41; Deborah Folsom, botanist at the Henry E. Huntington Library, San Marino, California, interview by author, 9 August 1991.

10. *Price List of Trumbull and Beebe, Fruit and Ornamental Trees* (San Francisco: Trumbull and Beebe, 1890–91), passim; Harry M. Butterfield, "Dates of Introduction of Trees and Shrubs to California" (manuscript produced by Landscape Horticulture, University of California at Davis, 1964).

11. "Few pests are indigenous," the authors write, "almost all have been brought in rather recently, along with the plants they attack." Ralph E. Smith and collaborators Harry S. Smith, Henry J. Quayle, and E. O. Essig, "Protecting Plants from Their Enemies," 239.

12. A. Freeman Mason, *Spraying, Dusting, and Fumigating of Plants: A Popular Handbook on Crop Protection* (New York: Macmillan, 1929), 300, 343.

13. O. E. Bremner, *Destructive Insects and Their Control* (Sacramento: Superintendent of State Printing, 1910); Alexander Craw, *Destructive Insects, Their Natural Enemies, Remedies, and Recommendations* (Sacramento: State Printing Office, 1891), passim; Smith and collaborators, "Protecting Plants," 245.

14. California Fruit Growers, *Report . . . 1884*, 5.

15. C. W. Woodworth, *Cooperative Work*, 188.

16. O. E. Bremner, *Destructive Insects*, 1.

17. California Fruit Growers, *Report of the First Annual Convention* (San Francisco: [Pacific Rural Press?], 1882), 2; Cleland and Hardy, *March of Industry*, 85; Howard Seftel quotes the same figures in "Government Regulation," 377.

18. Smith and collaborators, "Protecting Plants," 241.

19. O. E. Bremner, *Destructive Insects;* for other formulas, see Alexander Craw, *Destructive Insects.* Also see Robert Snetsinger, *The Ratcatcher's Child* (Cleveland, Ohio: Franzak and Foster, 1983), passim. The American botanist John Bartram used tobacco for this purpose in the 1740s.

20. C. W. Woodworth, "Remedies for Insects and Fungi," *University of California Agricultural Experiment Station Bulletin No. 115* (1898): 3.

21. O. E. Bremner, *Destructive Insects.*

22. Alexander Craw, *Destructive Insects*, 32.

23. C. H. Dwinell, "Entomology in the College of Agriculture," *University of California Agricultural Experiment Station Bulletin No. 16* (1884): 1.

24. Committee on Phylloxera, Vine Pests, and Diseases of the Vine for the State Board of Viticultural Commissioners, *Viticultural Map* (San Francisco: Britton and Rey, Lithographers, 1880). Fresno, which had not yet become a major raisin-producing region in 1880, was spared the phylloxera.

25. Eugene W. Hilgard, "Repression of the Phylloxera," in *University of California, College of Agriculture, Report of the Viticultural Work during the Seasons 1883–4 and 1884–5* (Sacramento: State Printing Office, 1886), 176–80.

26. Ibid., 179–84.

27. Ibid., 205.

28. Resistant rootstocks included Riparia Gloire de Montpellier and Riparia Grande Glabre, considered to be among the best. E. W. Hilgard and W. J. V. Osterhout, *Agriculture for Schools of the Pacific Slope* (New York: Macmillan, 1910), 312; Eugene W. Hilgard, *University of California, College of Agriculture, Report of the Viticultural Work during the Seasons 1885 and 1886* (Sacramento: State Printing Office, 1886), 139.

29. This chapter will use "codling" moth rather than "codlin" moth, though growers considered both spellings acceptable. California Fruit Growers, *Report... 1882*, 2.

30. Ibid.

31. C. H. Dwinelle in California Fruit Growers, *Report . . . 1882*, 7.

32. California State Commission of Horticulture, *Horticultural Statutes and County Ordinances of California in Force July 1, 1905* (Sacramento: Superintendent of State Printing, 1905), 13–15.

33. California Fruit Growers, *Report . . . 1884*, 7.

34. Frederick M. Maskew, *A Sketch of the Origin and Evolution of Quarantine Regulations* (Sacramento: California State Association of County Horticultural Commissioners, 1925), 7.

35. California State Commission of Horticulture, *Horticultural Statutes and County Ordinances*, 7.

36. Ibid., 7–9; Maskew, *Origin and Evolution of Quarantine Regulations*, 12–20, 35–37.

37. Fresno *Weekly Expositor*, 13 January 1886.

38. California Board of Horticultural Commissioners, *Meeting* (1882?), 14.

39. California Fruit Growers, *Report . . . 1882*, 2–3.

40. Howard Seftel notes, "The first move to involve government with the California fruit industry came not from angry consumers of wormy fruit. . . . It came, rather, from farsighted entrepreneurs who realized that the success of their enterprise hinged on producing a consistently healthy crop." Howard Seftel, "Government Regulation," 374; Gustavus A. Weber, *The Plant Quarantine and Control Administration: Its History, Activities, and Organization* (Washington, D.C.: Brookings Institution, 1930), 6–8; Smith and collaborators, "Protecting Plants," 242–49.

41. Smith and collaborators, "Protecting Plants," 249. The scale's scientific name is *Icerya purchasi*.

42. Cleland and Hardy, *March of Industry*, 102–3.

43. John Isaac, *Bug vs. Bug: Nature's Method of Controlling Injurious Species* (Sacramento: State Horticultural Commission, 1906), 5.

44. Ellwood Cooper, *Bug vs. Bug: Parasitology* (N.p., 1913), 9; Cleland and Hardy, *March of Industry*, 102–3; Alexander Craw, "What California Has Done for Horticulture," *Proceedings of the Twenty-Sixth Fruit-Growers' Convention . . . 1901* (Sacramento: Superintendent of State Printing, 1902), 160–65.

45. State Board of Horticulture, *Report on the Importation of Parasites*, 9; Smith, "Protecting Plants from Their Enemies," 251.

46. Isaac, *Bug vs. Bug*, 7.

47. Cooper, *Bug vs. Bug*, 6–8.

48. Koebele had no answers for how to manage a transition from chemicals

to predator insects, except to advocate the immediate importation of fifteen hundred species of ladybug. State Board of Horticulture, *Report on the Importation of Parasites*, 10–11. Bees also died from spraying, although proponents of insecticides refused to admit the fact for many years. James Whorton, *Before Silent Spring: Pesticides and Public Health in Pre-DDT America* (Princeton, N.J.: Princeton University Press, 1974), 28–29.

49. Cooper, *Bug vs. Bug*, 4–6.

50. Ibid.

51. California State Board of Horticulture, *Report on the Importation of Parasites and Predaceous Insects* (Sacramento: State Printing Office, 1892), 14.

52. Ibid.

53. [E. J. Wickson], "A Few Facts about Beneficial Insects," (1903), Records of the College of Agriculture and the Agricultural Extension Service, Farmers' Institute Letters, March 1903–October 1904, Bancroft Library.

54. Yet biological control survived. In 1907, the state of California established its own insectary at Sacramento to breed and release beneficials. The cryptolaemus ladybug attacked the citrus mealybug and other mealybugs and was bred by the millions in the 1920s. Harry Scott Smith encouraged research at the University into the 1950s. The release of sterile Mediterranean fruit flies over Los Angeles is a recent example of biological control. See Harry Smith and A. J. Basinger, "History of Biological Control," reprint from *California Cultivator*, 25 October 1947, Bancroft Library.

55. *Vedalia* 4 (May 1937).

56. C. W. Woodworth, "Cooperative Work in Economic Entomology," 188.

57. James Whorton, *Before Silent Spring*, 24.

58. Ibid., 20.

59. Ibid., 20; Professor A. E. Verrill of Yale University said of Paris green: "It may be questioned whether it is safe or advisable to mix dangerous mineral poisons with the soil, for the arsenic and copper will remain in the earth, or may be absorbed by growing vegetables, or cause mischief in other ways." *"Buhach" The California Insecticide*, pamphlet in the collection of the Bancroft, 11.

60. Whorton, *Before Silent Spring*, 20.

61. Lead arsenate is made by oxidizing arsenious oxide (white arsenic) to arsenic oxide and combining with lead acetate. $Pb_4PbOH (AsO_4)_3$. A. Freeman Mason, *Spraying, Dusting, and Fumigating of Plants* (New York: Macmillan, 1929), 40.

62. George E. Colby, "Arsenical Insecticides: Paris Green; Commercial Substitutes; Home-Made Arsenicals," *University of California Agricultural Experiment Station Bulletin No. 151* (1903), 4.

63. Leo Gardner, employee of California Spray-Chemical Company between 1923 and 1969, interview by author, 23 September 1992 and 8 February 1993 (the second by telephone).

64. Liberty Hyde Bailey, *The Horticulturist's Rule Book* (New York: Macmillan, 1899).

65. Smith and collaborators, "Protecting Plants," 263.

66. University of California Agricultural Experiment Station. *Publications of the Agricultural Experiment Station for the Period 1877–1918* (Berkeley: University of California, 1919), 2–4. This is a list of all published reports and bulletins.

67. Colby, "Arsenical Insecticides," 4.

68. C. W. Woodworth, "Cooperative Work in Economic Entomology," 188.

69. See Fitzgerald, *Business of Breeding*, 87; A. C. True, *A History of Agricultural Extension Work in the United States, 1785–1923* (Washington, D.C.: United States Department of Agriculture Miscellanies Publication 15, 1929), passim.

70. C. W. Woodworth, "Cooperative Work in Economic Entomology," 188.

71. Ibid. See University of California Agricultural Experiment Station, *Report of the Work of the Agricultural Experiment Station of the University of California from June 30, 1903, to June 30, 1904* (Sacramento: State Printing Office, 1904), 85; *Report of the Work of the Agricultural Experiment Station of the University of California from June 30, 1901, to June 30, 1903* (Sacramento: State Printing Office, 1903), 107.

72. Colby, "Arsenical Insecticides," 4.

73. *California Cultivator*, 20 (6 March 1903); California Fruit Growers, *Report of the Fifth Annual Convention of the California Fruit Growers* (San Francisco: Pacific Rural Press, 1886), 3.

74. Lime sulfur, lime sulfur and tobacco mixture, miscible oils, kerosene emulsions, and the arsenicals are examples of "kitchen formulas." H. J. Quayle, *University of California Agricultural Experiment Station Circular No. 66* (1911).

75. C. W. Woodworth, "List of Insecticide Dealers," *University of California Agricultural Experiment Station Circular No. 79* (July 1912); Whorton, *Before Silent Spring*, 22–23.

76. W. G. Klee, "The Woolly Aphis and Its Repression," *University of California Agricultural Experiment Station Bulletin No. 55* (1886).

77. California State Archives, Economic Poison Certificates, 1921–22, Sacramento, California.

78. Growers could perform a crude test in the orchard to assess the purity of Paris green. They placed drops of liquid on a mirror or glass and observed the track. A track with white specks indicated adulteration; a clear track meant that the sample was probably safe to use. But the first result might also mean improper mixing. And the test did not indicate the nature of the adulteration. *Pacific Rural Press* 59 (10 March 1900).

79. Colby, "Arsenical Insecticides," 12–17.

80. *Pacific Rural Press* 59 (10 March 1900).

81. Colby, "Arsenical Insecticides," 18, 3.

82. Ibid., 12–17.

83. Smith and collaborators, "Protecting Plants," 264.

84. *Pacific Rural Press* 60 (1 September 1900).

85. Hilgard and Osterhout, *Agriculture for Schools of the Pacific Slope*, 183; all Paris green sold in California had to contain at least 50 percent arsenious oxide and not more than 4 percent uncombined arsenious oxide. *An Act to Prevent Fraud in the Sale of Paris Green Used as an Insecticide*, reprinted in Colby, "Arsenical Insecticides."

86. Colby, "Arsenical Insecticides," 3.

87. Minutes of the State Board of Horticulture, 1883–1902, California State Archives.

88. More comprehensive than the first, the law of 1911 called for the prosecution of anyone caught manufacturing or selling adulterated or misbranded

insecticide. *The Insecticide Law, California Statutes of 1911, Chapter 653*, reprinted in C. W. Woodworth, *University of California Agricultural Experiment Station Circular No. 65* (July 1911).

89. C. W. Woodworth, "List of Insecticide Dealers," *University of California Agricultural Experiment Station Circular No. 79* (July 1912).

90. C. W. Woodworth, "The Insecticide Industries in California," *Journal of Economic Entomology* 5 (August 1912): 358.

91. Even as late as 1915, however, fraudulently manufactured insecticide circulated through California. *University of California Agricultural Experiment Station Circular No. 141* (1915).

92. C. W. Woodworth, "Cooperative Work in Economic Entomology," 187; *California Cultivator* 20 (16 January 1903).

93. University of California Agricultural Experiment Station, *Report of the Work of the Agricultural Experiment Station of the University of California from June 30, 1901, to June 30, 1903*, 107; Warren T. Clark to George B. Katzenstein, Manager of the Earl Fruit Company, 2 May 1906, Records of the College of Agriculture and the Agricultural Extension Service, Farmers' Institute Letters, March 1905 to August 1906, Bancroft Library; "California Spray Chemical Celebrates 20th Anniversary Tonight," unidentified newspaper article, perhaps from an in-house publication, perhaps from the Watsonville *Pajaronian*, 1927, Chevron Corporation Corporate Library, San Francisco.

94. Warren T. Clark to George B. Katzenstein, 2 May 1906, Farmers' Institute Letters, Bancroft; L. F. Czufin to E. W. Cannon, 13 April 1953, Chevron Corporate Library, San Francisco.

95. W. H. Volck to C. W. Woodworth, 13 July 1902, Correspondence Book, Charles W. Woodworth Papers, Bancroft Library.

96. Basic lead arsenate used a lead litharg plus a nonreactive catalyst to reduce the amount of soluble arsenic acid. The process tied the arsenic more tightly to the lead for a release of arsenic in the stomach of the insect rather than on the trees. "Basic" probably referred to the higher pH of the chemical as compared with older, more acidic lead arsenates. Leo Gardener, interview, 7 February 1993.

97. Leo Gardner, *The First Thirty Years: The Early History of the Company Now Known as Ortho Division, Chevron Chemical Company* ([San Francisco?]: Chevron Chemical, 1978), 1–17.

98. E. O. Essig, *Injurious and Beneficial Insects of California* ([Sacramento: State Commission of Horticulture, 1913?]), 318.

99. F. W. Braun came to Los Angeles early in the century to sell drugs and chemicals wholesale. "Men interested in developing fumigation brought their troubles to me. I did not know much about cyanide, and at that time I knew nothing whatsoever of fumigation, but I believed that I knew how to push the buttons and pull the strings so as to get not only the information but the goods." Braun Corporation, *School of Fumigation . . . Held at Pomona, California, August 9–13, 1915* (Los Angeles: Braun Corporation, 1915), 43 and passim.

100. Deborah Fitzgerald writes that the Illinois Experiment Station lost authority among farmers when private seed companies such as Funk Brothers began to produce hybrid corn. Both the university and the company saw themselves as research institutions in competition for "the attention and respect of Illinois farmers." Differences in circumstance as well as the nature of the product in

question distinguish the relationships between the University of California and insecticide companies. First, growers did not have to be told that insecticide was good for their productivity. No controversy existed at the experiment station about the benefits of insecticide. So companies that manufactured the chemicals according to university standards found no conflict with the authorities in Berkeley. Also, the university still maintained its authority by reviewing the products of the spray companies, informing growers about which worked best and warning them away from bogus products. Finally, the revolving door connecting college and company—as the story of Ortho illustrates—blurred the distinction between faculty and research staff. See Deborah Fitzgerald, *Business of Breeding*.

101. C. W. Woodworth, "The Insecticide Industries in California," *Journal of Economic Entomology* 4 (August 1912): 358.

102. Sherwin-Williams Company, *Spraying: A Profitable Investment* (Cleveland: In-house publication, [1918?]), 7–8.

103. Ibid., 19.

104. Richard L. Adams (1883–1957) was professor of agricultural economics at the College of Agriculture and the experiment station for over thirty years. Adams wrote of his findings that they present "practice in *commercial* production. They are not designed to indicate what should be done, but rather what is being done by men specializing in these crops." R. L. Adams, *Farm Management* ([Berkeley]: [1918?]), 83 and *Farm Management Notes (for California), 7th ed.* (Berkeley: Associated Students Store, 1921), 72; G. H. Miller, "Operating Costs of a Well-Established New York Apple Orchard," *United States Department of Agriculture Bulletin No. 130* (21 August 1914); C. R. Crosby, "Cost of Dusting and Spraying a New York Apple Orchard," *Journal of Economic Entomology* 3 (June 1916); E. J. Wickson, *California Fruits*, 229.

105. In 1909, E. J. Wickson estimated the average gross returns of deciduous fruits at $88.00 per acre, but that $750.00 per acre was the upper extreme. E. J. Wickson, "Cost of Planting and Care of Trees and Vines to Bearing Age," manuscript from the Division of Rural Institutions, Giannini Foundation of Agricultural Economics Library, Berkeley, California.

106. Compiled from R. L. Adams, *Farm Management*, 83, preface; and *Farm Management Notes*, 72–73, preface.

107. Gross profit on one acre of pears: two tons of fresh pears equals eighty-seven boxes (forty-six pounds each),which, at $1.25, amounts to $108.75. Two tons of canned pears selling at $30.00 per ton amounts to $60.00. And one ton of dried pears at $.08 a pound comes to $160.00. All these are "usual" prices and quantities. Ibid.

108. Woodworth, "Quantitative Entomology," *Annals of the Entomological Society of America*, 8 (December 1915): 381–83. A. Freeman Mason concluded, "Fifty years ago, spraying was practically unknown; now it is universally recognized as essential to success." A. Freeman Mason, *Spraying, Dusting and Fumigating of Plants*, 12; Also see Samuel T. Maynard, *Successful Fruit Culture: A Practical Guide to the Cultivation and Propagation of Fruits* (New York: Orange and Judd, 1905; reprint, 1913), 220.

109. C. W. Woodworth, "Remedies for Insects and Fungi," 3. My emphasis.

110. C. W. Woodworth, "Directions for Spraying the Codling-Moth," *University of California Agricultural Experiment Station Bulletin No. 155* (1904), 6–15.

111. Ibid., 16

112. *Pacific Rural Press* 59 (10 March 1900).

113. Ruth deForest Lamb, *American Chamber of Horrors: The Truth about Food and Drugs* (New York: Farrar and Rinehart, 1936), 196; Whorton, *Before Silent Spring*, 95. An English study of American fruit conducted in 1926 warned that arsenic on the skin of sprayed fruit posed a risk to young children and susceptible adults. H. E. Cox, "The Occurrence of Arsenic in Apples," *Analyst* 51 (March 1926): 132. Lead was a very common environmental poison in early twentieth-century America, showing up in paints, canned goods, metal containers, cooking utensils, water pipes, and exhaust gasses from automobiles. James Whorton, *Before Silent Spring*, 176–211.

114. Whorton, *Before Silent Spring*, 96.

115. The program Swett probably referred to was the removal of spray residues from pears by wiping them in the packinghouse. Frank T. Swett, *The Pear Growers' Progress: Address of Frank T. Swett* (San Francisco: California State Pear Growers' Association, 1920).

116. "Brethren, Let Us Spray," *Pacific Rural Press* 99 (24 April 1920).

117. This discussion can only be suggestive of a topic that is covered more fully by James Whorton in *Before Silent Spring*, 176–212.

118. C. N. Meyer, Binford Throne, Florence Gustafson, and Jerome Kingsbury, "Significance and Danger of Spray Residue," *Industrial and Engineering Chemistry* 25 (June 1933): 624.

119. Ibid., 625.

120. *California Cultivator* 33 (8 July 1909). As Whorton points out, few people who ate sprayed fruit and reacted to it knew the reason for their illness. Children were especially vulnerable to poisoning. Whorton, *Before Silent Spring*, 95.

121. W. B. White, "Poisonous Spray Residues on Vegetables," *Industrial and Engineering Chemistry* 25 (June 1933): 621–22.

122. Ibid.; H. E. Cox, "The Occurrence of Arsenic in Apples," *The Analyst* (March 1926): 132–33.

123. "All Honor to These Growers," *Pacific Rural Press* 112 (24 July 1926).

124. Ibid., 125.

125. William Haedden, *Colorado Agricultural Experiment Station Bulletin No. 131*, quoted in Haedden, "Arsenical Poisoning of Fruit Trees," *Journal of Economic Entomology* 2 (June 1909): 242.

126. Ibid.

127. E. D. Ball, "Is Arsenical Spraying Killing Our Fruit Trees?" *Journal of Economic Entomology* 2 (April 1909): 142–48.

128. E. D. Ball, E. G. Titus, and J. E. Greaves, "The Season's Work on Arsenical Poisoning of Fruit Trees," *Journal of Economic Entomology* 3 (April 1910): 187–97.

129. Plants absorb arsenic as a normal process of uptake from the soil. Arsenical residues in the soil increase the level found in plants, but only by very small amounts. The highest level recorded was 5.2 ppm—a very high level compared to

an average of 1.5 ppm found in other studies. Gary R. Sandberg and Ingrid K. Allen, "A Proposed Arsenic Cycle in an Agronomic Ecosystem," in *Arsenical Pesticides*, edited by E. A. Woolson, American Chemical Society Symposium Series 7 (Washington, D.C.: American Chemical Society, 1975), 135–39. No study found a higher concentration than 2 ppm, and this in an orchard where the concentration of arsenic in the soil read 233 ppm. W. D. Guenzil, ed., *Pesticides in Soil and Water* (Madison, Wis.: Soil Science Society of America, 1974), 208–9.

130. S. C. Vandecaveye, G. M. Horner, and C. M. Keaton, "Unproductiveness of Certain Orchard Soils as Related to Lead Arsenate Spray Accumulations," *Soil Science* 42 (1936): 203–13.

131. Sandberg and Allen, "A Proposed Arsenic Cycle," 124. I should point out, however, that the ecology of orchards has another side: they need no heavy cultivation and offer a possible method of agriculture without soil erosion or topsoil loss. At least one geographer in the past recognized in tree crops the salvation of American agriculture with its wasteful soil practices. See J. Russell Smith, *Tree Crops: A Permanent Agriculture* (New York: Harcourt Brace and Company, 1929). The book includes a bibliography on soil erosion. Also see Colin A. M. Duncan, *Centrality of Agriculture*, 155.

132. Cleland and Hardy, *March of Industry*, 97.

133. *Pacific Rural Press* 111 (27 February 1926).

134. George Hecke to Frank Honeywell, editor of the *Pacific Rural Press*, 18 June 1926, Frank Honeywell Papers: 1926–1931, Bancroft.

135. George Hecke, *The Mediterranean Fruit Fly: A Fruit Enemy Which Must Be Kept Out of California* (Sacramento: State Department of Agriculture, 1929).

136. Ibid.

137. *California Cultivator* 20 (9 January 1903).

138. Leo Gardener, *First Thirty Years*, 39–41. The list of companies that emerged during this time to produce pesticides includes Shell Chemical Company, Union Carbide Corporation, Dow Chemical Company, Allied Chemical, Velsicol, and Chemagro Corporation.

5.   White Men and Cheap Labor

1. Liberty Hyde Bailey understood that industrial agriculture would change the character of rural work: "[W]e must face the fact, however, that a necessary result of the organization of country life and the specialization of its industries . . . will be the production of a laboring class by itself." *The Country-Life Movement*, 144. The labor system I describe in this chapter did not differ substantially from that for crops like vegetables and cotton. Although I use fruit as my primary example, migrant labor became general to specialized agriculture all over California. This chapter assumes a slightly broader conception than the previous chapters.

2. Many preceding studies of California agriculture have informed this chapter, especially Cletus Daniel, *Bitter Harvest: A History of California Farmworkers, 1870–1941* (Ithaca: Cornell University Press, 1981); Sucheng Chan, *This Bitter-Sweet Soil*; Linda C. Majka and Theo J. Majka, *Farm Workers Agribusiness and the State* (Philadelphia: Temple University Press, 1982); Gilbert González,

Labor and Community: Mexican Citrus Worker Villages in a Southern California County, 1900 –1950 (Urbana and Chicago: University of Illinois Press, 1994); Camille Guerin-Gonzales, *Mexican Workers and American Dreams* (New Brunswick, N.J.: Rutgers University Press, 1994); Ruiz, *Cannery Women/Cannery Lives*. Also refer to Paul S. Taylor, "Foundations of California Rural Society," *California Historical Society Quarterly* 24 (1945): 193–228 and Paul S. Taylor and Tom Vasey, "Historical Background of California Farm Labor," *Rural Sociology* 1 (September 1936); Carey McWilliams, *Factories in the Field: The Story of Migratory Farm Labor in California* (Boston: Little, Brown and Company, 1939); Varden Fuller, "The Supply of Agricultural Labor as a Factor in the Evolution of Farm Organization in California," United States Congress, Senate Committee on Education and Labor, Hearings Pursuant to Senate Resolution 266, Exhibit 8762-A, 76th Congress, 3rd Session, 1940, 19777–898; Lloyd Fisher, *The Harvest Labor Market in California* (Cambridge: Harvard University Press, 1953); Carleton H. Parker, *The Casual Laborer and Other Essays*, edited by Cornelia Stratton Parker (New York: Harcourt, Brace, and Howe, 1920).

3. For a sample of this debate see H. C. Bennett, *Chinese Labor;* and *Minority Report from Committee on Chinese* (3 December 1878), manuscript report in the Records of the Secretary of State, Miscellaneous Constitutional Convention Proceedings, California State Archives. For scholarly treatments see Gunther Paul Barth, *Bitter Strength: A History of the Chinese in the United States, 1850 – 1870* (Cambridge: Harvard University Press, 1964) and Alexander Saxton, *The Indispensable Enemy: Labor and the Anti-Chinese Movement in California* (Berkeley and Los Angeles: University of California Press, 1971).

Tomás Almaguer argues in *Racial Fault Lines* that race served as "the central organizing principle of hierarchical group relations in California." Placed in the context of Almaguer's work, the growers' term "cheap labor" signifies this privileged position and the continuing racialized struggle. Yet the centrality of race does not explain the relationship between white growers and the white workers who arrived during the 1930s. Growers exploited them without appealing to racial supremacy. *Racial Fault Lines*, 3, 209.

4. Lawanda F. Cox, "The American Agricultural Wage Earner, 1865–1900: The Emergence of a Modern Labor Problem," *Agricultural History* 22 (April 1948): 100–103.

5. R. L. Adams and T. R. Kelly, "A Study in Farm Labor," *University of California Agricultural Experiment Station Circular No. 193* (March 1918): 7.

6. Horace Bell, *Reminiscence of a Ranger* (Los Angeles: n.p., 1881), 24–36, quoted in John Caughey and LaRee Caughey, eds., *Los Angeles: Biography of a City* (Berkeley and Los Angeles: University of California Press, 1976), 124–26.

7. William Smythe, *Conquest of Arid America*, 145; Donald Pisani, "Panacea or Curse: Attitudes toward Irrigation in Nineteenth-Century California," in *From Family Farm*, passim.

8. Edward James Wickson, *California Fruits*, 335–69.

9. Winkler, *General Viticulture*, 333, 338. An acre-inch is an acre of land flooded to the depth of one inch. The capacity of an irrigation canal is determined by how many acre-feet of water cross a specified point every second.

10. Winkler, *General Viticulture*, 565; "Imperial Fresno," 13; Wickson, *California Fruits*, 355–69.

11. Gustav Eisen, *Raisin Industry*, 180; R. L. Adams, *Farm Management*.

12. *Complete Practical Farmer* (New York: n.p., 1839), 71, quoted in Clarence H. Danhof, *Change in Agriculture*, 137.

13. Ibid., 141.

14. Warren, *Farm Management*, 107.

15. W. J. Spillman, "Seasonal Distribution of Labor on the Farm," *Yearbook of the United States Department of Agriculture, 1911* (Washington: Government Printing Office, 1912), 270.

16. Ibid., 272–73.

17. Chan, *This Bitter-Sweet Soil*, 41.

18. This argument is similar to the one made by Varden Fuller in his 1939 dissertation, "The Supply of Agricultural Labor as a Factor in the Evolution of Farm Organization in California." Fuller argues that large farms emerged in regions with heterogeneous populations in which a privileged class gained access to credit and entrepreneurial skill and gave the underprivileged class no other choice but to serve as labor. He asserts that landowners converted wheat fields into fruit orchards on the same large scale and that the presence of potential workers like Chinese immigrants made this transition possible.

But labor was only one factor that influenced the size of orchards. If fruit could be cultivated on the same scale as wheat, why did so many land monopolists decide to subdivide and sell off their acres? Fuller does not consider that fruit made all kinds of demands on landowners that grain never did. Few people in the 1880s and 1890s had any experience with growing fruits or vegetables on a commercial scale. After landowners worked out problems of labor and insects and distribution, they consolidated those twenty-, forty- and eighty-acre parcels into thousand-acre units—the units that Fuller saw dominating the countryside in 1939. Varden Fuller, "The Supply of Agricultural Labor," 19777–898; Chan, "Working Hands," in *This Bitter-Sweet Soil*, 279, passim.

19. Warren, *Farm Management*, 95–99.

20. Frank Adams, "Third Talk—Demonstration Train," Frank Adams Papers, folder 146, WRCA. Published as "The Agricultural Situation in California," *University of California Agricultural Experiment Station, Agricultural Extension Service Circular 18* (1928).

21. Paul S. Taylor, "Patterns of Agricultural Labor Migration within California," *United States Department of Labor, Bureau of Labor Statistics, Serial No. 840* (Washington, D.C.: Government Printing Office, 1938), 1–2. Taylor points out, however, that these are estimates for working laborers and do not account for the unemployed. R. L. Adams and T. R. Kelly, "Study in Farm Labor," 42–43. The authors figured *maximum* labor needs for one hundred acres: hops, 300 people for three to four weeks; prunes, 10 for four to six weeks; raisins, 20 for three to four weeks; pears, 100 for four to six weeks; J. D. Culbertson, "Housing of Ranch Labor," *First Annual Report of the California Citrus Institute* (San Bernardino: California Citrus Institute, 1920), 97–99.

22. See chapters 2 and 6 for some of the reasons behind large-acreage production.

23. Fisher, *Harvest Labor Market*, 2; J. Karl Lee, "Economies of Scale." For an opposing view, Paul Taylor maintained that 69.8 percent of all farms in California employed no wage labor (85.9 percent for the United States as a whole).

Taylor did not specify the time of year or the types of farms. Paul S. Taylor, "Factors Which Underlie the Infringement of Civil Rights in Industrial Agriculture," in *Violations of Free Speech and Rights of Labor*, Pt. 62, Exhibit 9573 (Washington D.C.: Government Printing Office, 1940), 22488–922.

24. Oliver E. Baker, "Agricultural Regions," 295–96; According to the California Citrus Institute, one person could do all the work necessary on a diversified 40.5 acre farm; one person could take care of 135 acres planted in wheat; but in California one person could tend to only *two to three acres* in lemons, pick them, and pack them for shipment. *First Annual Report of the California Citrus Institute* (San Bernardino: California Citrus Institute, 1920), 98.

25. A. B. Butler, "California and Spain" in *California—A Journal of Rural Industry* 3 (15 March 1890).

26. H. P. Stabler, "The California Fruit-Grower, and the Labor Supply," *Transactions of the Twenty-Seventh State Fruit-Growers' Convention . . . 1902* (Sacramento: Superintendent of State Printing, [1903]), 268.

27. George Hecke, "The Pacific Coast Labor Question from the Standpoint of a Horticulturist," *Proceedings of the Thirty-Third State Fruit Growers' Convention . . . 1907*, 67.

28. Ibid., 71.

29. William H. Mills, "Annual Address," *Transactions of the California State Agricultural Society* (Sacramento: State Printing Office, 1891), 200; Cletus Daniel, *Bitter Harvest*, 45; Lloyd H. Fisher, *Harvest Labor Market*, 11.

30. California Development Association, *Survey of the Mexican Labor Problem in California* ([San Francisco]: California Development Association, [1928]), 8. The board of directors included Clause Spreckles, the sugar beet grower.

31. Ibid.

32. Chan, *This Bitter-Sweet Soil*, 2–3, 31; *Minority Report from* [sic] *Committee on Chinese* (3 December 1878), manuscript report in the Records of the Secretary of State, Miscellaneous Constitutional Convention Proceedings, California State Archives. One member of the state committee said of the Chinese worker: "He is not an emigrant. He is not eligible to [*sic*] American citizenship. He is imported as a chattel."

33. Cletus E. Daniel makes a similar argument. Daniel states that a lingering agrarianism influenced California growers and farmers to seek workers like themselves, and that the appearance of a racially oppressed workforce coincided with the erosion of agrarian idealism. The trouble with the argument is that it asks the reader to believe that farmers in California once took agrarian ideology seriously. Daniel admits that no rural tradition existed in California comparable to that of the older states. It seems more likely that certain farmers romanticized an earlier farmer's republic, especially when they confronted land monopoly. For a discussion of the rhetoric of agrarianism and how it slipped into romanticism with the decline of the preindustrial countryside, see Leo Marx, *Machine in the Garden;* and Raymond Williams, *The Country and the City* (New York: Oxford University Press, 1973); Daniel, "The Erosion of Agrarian Idealism," chapter 1 in *Bitter Harvest*.

34. John Summerfield Enos, Report, *Second Biennial Report of the Bureau of Labor Statistics of the State of California for the Years 1885–1886* (Sacramento: Superintendent of State Printing, 1887), 61.

35. Chinese taught their employers how to pack fruit for shipment, how to prune trees to bring forth the most fruit, and in general how to manage a clean garden. The Chinese became successful truck gardeners all over the state, especially near major towns and cities and in the San Joaquin River Delta. In the southern San Joaquin Valley, they worked as tenant farmers and migrant labors. Sucheng Chan, *This Bitter-Sweet Soil*, 235, 334.

36. Hugh LaRue, "Opening Address," *Transactions of the California State Agricultural Society* (Sacramento: State Printing Office, 1883), 20.

37. Ibid.

38. Ibid.

39. Ibid., 18–19.

40. *Second Biennial Report of the Bureau of Labor Statistics of the State of California for the Years 1885–1886* (Sacramento: Superintendent of State Printing, 1887), 47.

41. Ibid., 60.

42. Ibid., 64; *Fresno Weekly Expositor*, 7 February 1886.

43. *Maywood Colony Advocate* 1 (1 November 1899); *California—A Journal of Rural Industry*, 14 (12 April 1890); C. A. Rohrabacher, *Fresno County, California*, 47–48; *Pacific Rural Press* 45 (10 June 1893); Account books, R. B. Blowers Papers, Carton 5, Bancroft Library, University of California at Berkeley.

44. H. P. Stabler, "California Fruit-Grower," 269.

45. George Hecke, "Pacific Coast Labor Question," 67.

46. In one ill-advised plan, emissaries from California proposed to travel to France, there to contract with entire villages from the southern grape-growing regions and move them to Fresno County. The French country people already knew the crop, reasoned the plan's supporters. Nothing came of it. Note that the French peasants combined two characteristics attractive to the growers: the appearance of docility and European lineage. *Fresno Weekly Expositor*, 7 February 1886.

47. H. P. Stabler, "California Fruit-Grower," 270; *Fresno Morning Republican*, 1 January 1896.

48. H. P. Stabler, "California Fruit-Grower," 271.

49. "Report of Committee on Farm Labor," in *Synopsis of the Proceedings of the Twenty-Eighth State Fruit-Growers' Convention . . . 1903* (Sacramento: Superintendent of State Printing, 1903), 89–91.

50. California Fruit Growers, "Report of the Committee on Labor," in *Official Report of the Twenty-Ninth Fruit-Growers' Convention . . . December 1903* (Sacramento: Superintendent of State Printing, 1904), 217–21.

51. California Fruit Growers, *Transactions . . . 1902*, 280–81; Daniel, *Bitter Harvest*, 53–55.

52. Ibid. See Stabler's paper and the discussion that followed.

53. Cletus Daniel states on this subject: "When farm employers began in the 1880s to adopt the view that white workers constituted the ultimate solution to their labor problems, they neither contemplated nor desired changes in the highly commercial structure of the state's agricultural industry or in the basic relationship between themselves and the dependent wage earners who labored for them." Daniel, *Bitter Harvest*, 47.

54. Ibid., 276.

55. California Fruit Growers, *Proceedings of the Fiftieth State Fruit Growers Convention, 1917* (Sacramento: State Printing Office, 1918), 78. Field crops often required hoeing to thin out the plants.

56. Gilbert González points out an important exception to the impermanence of labor housing. By the 1930s Mexican citrus workers in Orange County lived in villages, or *colonias*, and harvested throughout most of the year. González, *Labor and Community*, 60–66.

57. Commission of Immigration and Housing, *First Annual Report of the Commission of Immigration and Housing of California* (Sacramento: State Printing Office, 1915), 33.

58. Adams and Kelly, "Study of Farm Labor," 11–13.

59. California Fruit Growers, *Proceedings of the Thirty-Third Fruit Growers' Convention . . . 1907* (Sacramento: Superintendent of State Printing, 1908), 72.

60. *Pacific Rural Press* 99 (26 June 1920).

61. J. D. Culbertson, "Housing of Ranch Labor," in *First Annual Report of the California Citrus Institute* (San Bernardino: California Citrus Institute, 1920), 98.

62. Commission on Immigration and Housing, *Report on Unemployment to his Excellency Governor Hiram W. Johnson* (Sacramento: State Printing Office, 1914), 9.

63. Fisher, *Harvest Labor Market*, 9–10.

64. Commission on Immigration and Housing, *Report on Unemployment*, 9.

65. California Bureau of Labor Statistics, *Private Employment Agency Law* (San Francisco: State Printing Office, 1924); Seth R. Brown, *Report of the Division of State Employment Agencies of the Department of Industrial Relations of the State of California* (Sacramento: State Printing Office, 1928).

66. Seth R. Brown, *Report of the Division of State Employment Agencies*.

67. Adams and Kelly, "Study of Farm Labor," 30–33; Oliver Baker, "Agricultural Regions of North America," (April 1930), 167–90, continued in (July 1930), 299. White farm labor earned $2.25 to $3.50 for a nine-hour day. Board offered by the farmer usually cost $1.00 to $1.50 subtracted from wages each day. Harvest hands made $2.50 to $4.00 for nine hours. "Agricultural Wages," *Pacific Rural Press* 102 (9 July 1921). Wages paid for picking grapes were often determined on a per unit piece rate or as a flat daily rate. In 1889 picking earned $1.50 per day without board. By 1913 the piecework method seems to have been established. Examples:

1913 : $.02.5 per tray for raisin varieties in Fresno.
1914: same.
1915 : $2.17 per day without board.
1917: $.03 per tray for muscats and $.02.5 for Thompson seedless.
1918: $3.50 per day without board

Note the higher wages paid during the war. Into the 1930s wages were still $.02 cents per tray for seedless. Federal Writers Project, "Wage Chart by Crop: State of California, 1865–1938," part of "A Documentary History of Migratory Farm Labor in California," Papers of the Federal Writers Project of Oakland, California, Bancroft Library.

68. George Hecke, "Pacific Coast Labor Question," 68. Hecke's emphasis.

69. John Summerfield Enos, *Statement of John Summerfield Enos to the California State Horticultural Society* (Sacramento: State Printing Office, 1886), 16; California Fruit Growers, *Proceedings, 1917*, 78.

70. Irish continued, "He works a month, takes his $40, goes to town, throws it down his throat, and that is the last of him." John P. Irish, "Labor in the Rural Industries of California," in California Fruit Growers, *Proceedings . . . December 1907*, 64.

71. California Fruit Growers, *Proceedings of the Sixtieth Convention of Fruit Growers and Farmers . . . 1927* (Sacramento: State Printing Office, 1928), 119; California Fruit Growers, *Transactions, 1902*, 274.

72. California Fruit Growers, *Proceedings . . . 1917*, 86.

73. A. J. Pillsbury, "What Legislation Has Done, Is Doing, and Should Do for Agriculture," *Transactions of the Commonwealth Club of California* 6 (November, 1911): 428–30.

74. Thomas Forsyth Hunt, *How Can a Young Man Become a Farmer*, (printed address read before the San Jose Meeting of the Patrons of Husbandry of California, 21 October 1913 and at the University Farm, Davis, California, 4 December 1913), 4.

75. Harris Weinstock, before the Fruit Growers' Convention of 1914, quoted in Commission of Immigration and Housing, *Report on Unemployment*, 22.

76. California Fruit Growers, *Proceedings, 1917*, 78.

77. California Fruit Growers, *Report, 1907*, 58.

78. Ibid.

79. Ibid.

80. Tomás Almaguer, "Racial Domination and Class Conflict in Capitalist Agriculture: The Oxnard Sugar Beet Workers' Strike of 1903," *Labor History* 25 (summer 1984) and *Racial Fault Lines*, 184–204.

81. California Fruit Growers, *Report . . . 1907*, 69; Federal Writers' Project of Oakland, "The Contract Labor System in California Agriculture," Manuscript Collection of the Bancroft Library, [1938?], 20–29; Commonwealth Club of California, *Transactions* 13 (May 1918): 91–92; California Bureau of Labor Statistics, *Biennial Report, 1895–96*, 102, quoted in Cox, "American Agricultural Wage Earner," 102. By 1906 there were 41,000 Japanese in California.

82. California Fruit Growers, *Report . . . 1907*, 70; ibid., 92.

83. Letter, John Irish to the *Fresno Republican*, [1910], box 1, folder 3, John P. Irish Papers, Stanford University, Green Library, Special Collections.

84. John P. Irish, "The Japanese in California: When They Came," manuscript article (n.d. but 1920s), John P. Irish Papers, Box 3, folder 28, Stanford University, Green Library, Special Collections.

85. John P. Irish, "California and the Japanese," Manuscript article, (n.d. but 1910s), John Irish Papers, box 3, folder 28, Stanford University, Green Library, Special Collections.

86. Ibid.

87. Ibid.

88. While not excusing the IWW organizers, the [Marysville] *Democrat* (6 August 1913) blamed the violence on humiliating conditions of work, especially on the piecework wage of $.90 for a hundred-pound sack—$.10 below the going rate. Don Mitchell, *Lie of the Land: Migrant Workers and the California Landscape*

(Minneapolis: University of Minnesota Press, 1996), 36–40. For another account of the riot see Carey McWilliams, *Factories in the Field*, 152–67.

89. [Marysville] *Democrat*, 6 August 1913, in *Scrapbook on the Wheatland Riot*, vol. 1, Bancroft Library; Parker, "Report," 176–77.

90. Mitchell, *Lie of the Land*, 50–57. Mitchell shows how the California Commission on Immigration and Housing (CCIH) under Simon Lubin and George Bell developed a plan to eliminate the IWW in California and the Far West, how they gathered the support of western governors and the federal government, and how they presided over the surveillance and repression of IWW leaders, all between 1917 and 1919. As Mitchell puts it, "The end goal of the CCIH program was to immobilize the radically mobile and to replace them with a labor supply incapable of resistance." Mitchell, *Lie of the Land*, 76.

91. California Fruit Growers, *Proceedings of the Forty-Fifth Fruit Growers' Convention . . . 1914* (Sacramento: State Commissioner of Horticulture, 1915), 125.

92. George Hecke, "Argument against the Eight Hour Law," in *Amendments to the Constitution and Proposed Statutes with Arguments Respecting the Same* (Sacramento: State Printing Office, 1914), 58–59.

93. Ibid.

94. *Farmers News*, November 1914.

95. E. E. Bowles, "The Farm Labor Situation," in *Transactions of the Commonwealth Club of California* 13 (May 1918): 98.

96. David Vaught demonstrates this point well. George Pierce secured a crew of Sikhs from a State Employment Bureau at $.30 an hour. But the workers soon asked for $.05 more, and Pierce paid it. David Vaught, "'An Orchardist's Point of View': Harvest Labor Relations on a California Almond Ranch, 1892–1921," *Agricultural History* 69 (fall 1995): 584–85.

97. Richard Adams and T. R. Kelly, "Study in Farm Labor," 2.

98. Ibid.

99. George W. Pierce, "Farm Labor from an Orchardist's Point of View," in Commonwealth Club of California, *Transactions* 13 (May 1918): 88. Quoted in Vaught, "'An Orchardist's Point of View.'" 587.

100. Ibid., 33.

101. Ibid., 15; notes from a meeting held at University of California Berkeley on 25 June 1917, in the Herbert C. Jones Papers (1903–54), Stanford University, Green Library, Special Collections. Charles E. Warren of the Santa Clara Valley negotiated with an Oakland high school for boys to harvest the region's summer crops. He learned that an unspecified number of them would cost the growers $1,000. Warren called this paying "a premium on unskilled labor" and refused the offer. *Transactions of the Commonwealth Club of California* 23 (May 1918): 114–15.

102. Alice Prescott Smith, "The Battalion of Life: Our Woman's Land Army and Its Work in the West, *Sunset* 41 (November 1918): 30–33; Susan Minor, "Sisters All," *Overland Monthly* 73 (May 1919). Minor states that the battalion was founded in February of 1918 after a similar experiment the previous year in the state of New York. In the fall of 1918 it affiliated with the United States Department of Labor. Address, George Clements of the Agriculture Department of the Los Angeles Chamber of Commerce, 29 November 1918, "Articles and Papers," vol. 1, George Clements Papers, Carton 1, Bancroft Library, University

of California, Berkeley. Also see R. L. Adams, "The Farm Labor Problem," *The University of California Chronicle* 22 (April 1920): 200–16.

103. Smith, "The Battalion of Life."

104. California Fruit Growers, *Proceedings . . . 1914*, 125.

105. Adams and Kelly, "Study in Farm Labor," 15; California Fruit Growers, *Proceedings . . . 1917*, 77–78.

106. M. F. Tarpey, "Some Possibilities of the Development of New Labor during the War," California Fruit Growers, *Proceedings . . . 1917*, 78.

107. California Fruit Growers, *Proceedings . . . 1917*, 102.

108. In order to institute the program the secretary of labor, William B. Wilson, waived the literacy test and head tax provisions in the Immigration Act of 1917 after growers complained of shortages. All Mexicans were called "white" to satisfy the Act's stipulation that no nonwhite workers be allowed to immigrate. The program lasted until 1921. By that time 72,000 Mexican workers had crossed the border. According to Linda and Theo Majka, growers claimed that 80 percent of the Mexicans in the United States were subject to deportation. According to Camille Guerin-Gonzales, most returned home after working, but 40 to 60 percent remained in the United States. The government program predicted the Braceros Program of World War II. Majka and Majka, *Farm Workers*, 61–68; Guerin-Gonzales, *Mexican Workers*, 41–47.

109. Guerin-Gonzales, *Mexican Workers*, 41–47.

110. "Labor Conditions," *Monthly Labor Review* 11 (November 1920): 221–23. Most of the Mexican workers did not work in fruit, however; they concentrated in field crops like sugar beets and cotton.

111. Kathryn Cramp, Louise F. Shields, Charles A. Thomson, "Study of the Mexican Population in Imperial Valley, California," manuscript, "Made for the Committee on Farm and Cannery Migrants, Council of Women for Home Missions, 156 Fifth Avenue, New York City, March 31–April 9, 1926," Bancroft Library. Mexican farm workers tended to travel in families. Citrus growers sometimes offered to build them homes as a way to keep workers nearby. For their part, the Mexican families desired permanence. González, *Labor and Community*, 36.

112. Paul S. Taylor, *Patterns of Agricultural Labor Migration within California*, Serial No. 840, United States Department of Labor, Bureau of Labor Statistics. (Washington, D.C.: Government Printing Office, 1938), 3. The use of schoolchildren in this example provides a contrast to the use of white children from Oakland during the war.

113. Ibid., 6.

114. Cramp, Shields, and Thomson, "Study of the Mexican Population."

115. California Development Association, *Survey of the Mexican Labor Problem*, 3.

116. A. J. Pillsbury, "What Legislation Has Done," 428.

117. California Development Association, *Survey of the Mexican Labor Problem*, 8.

118. The bill amended the Immigration Act of 1924 by making its quota provision "applicable to Mexico, Cuba, Canada, and the countries of continental America and adjacent islands." United States Congress, House of Representatives, Committee on Immigration and Naturalization. 69th Cong., 1st Sess.,

"*Seasonal Agricultural Laborers from Mexico*," Hearings pursuant to HR 6741, 28 January 1926 (Washington, D.C.: Government Printing Office, 1926), 1. The Majkas report more on the story. The American Federation of Labor and small farmers wanted restrictions on Mexican immigration. And Box and others on the committee represented Southeastern cotton growers who sought to maintain their region's competitive advantage over Southwestern cotton growers. *Farm Workers*, 64.

119. United States Congress, House of Representatives, Committee on Immigration and Naturalization. *Seasonal Agricultural Laborers from Mexico*, 24.

120. *Pacific Rural Press* 111 (3 April 1926).

121. *Transactions of the Commonwealth Club of California* 13 (May 1918): 112–15; California Fruit Growers, *Proceedings . . . 1927*, 116. Lubin quoted from a public hearing held "some years ago" at the bequest of the United States Department of Labor in Los Angeles. In another example of how the state worked to regulate immigrant labor, Mexico, acting through its consulate, encouraged so-called conservative labor unions as a vehicle to resolve strikes in the growers' favor. Gilbert G. González, "Company Unions, the Mexican Consulate, and the Imperial Valley Agricultural Strikes, 1928–1934 *Western Historical Quarterly* 27 (spring 1996): 53–73.

122. California Fruit Growers, *Proceedings . . . 1927*, 118.

123. *Pacific Rural Press* 102 (13 August 1921); *Pacific Rural Press* 102 (12 November 1921).

124. Carey McWilliams suggests that the appearance of whites marked a turning point: "The entry of white workers into the fields symbolized the industrial maturity of California agriculture." *Factories in the Field*, 199.

6.   Natural Advantages in the National Interest

1. Herbert Hoover, *The Future Development of the Great Central Valley of California* (published address delivered before the Sacramento Chamber of Commerce, 27 June, 1925).

2. Oliver Baker, "Population Trends in Relation to Land Utilization," *Proceedings of the International Conference of Agricultural Economics II*, reprint, [1929], 295–306.

3. Ibid.; Wheeler McMillen, *Too Many Farmers: The Story of What Is Here and Ahead in Agriculture* (New York: William Morrow and Company, 1929), 12. It is difficult to confirm McMillen's figure that 4 million people left the occupation of farming. The census shows a decline from 1910 to 1930 of 2,175,145 people in all aspects of agriculture, including forestry—or from 12,659,082 people in 1910 to 10,483,917 people in 1930. The number of farmers actually increased slightly over the same period, though their portion among the working population fell. *Fourteenth Census of the United States: 1920, Occupations, vol. IV* (Washington D.C.: Government Printing Office, 1922), 35; *Fifteenth Census of the United States: 1930, Population, Occupations—General Report* vol. V, 412.

4. Oliver Baker, "Changes in Production and Consumption of Our Farm Products and the Trend in Population: Do We Need More Farm Land?" *The Annals of the American Academy of Political and Social Science* 142 (March 1929): 98.

5. Arthur Capper, *The Agriculture Bloc* (New York: Harcourt, Brace, and Company, 1922), 38.

6. Ibid., 11.

7. James Shideler, *Farm Crisis*, passim.

8. "California's 'Permanent Wave' of Prosperity," *Literary Digest* 80 (29 March 1924): 62–63; Samuel Fortier, "Agriculture and the Agricultural Engineer," *Agricultural Engineering* 4 (April 1923): 55–56.

9. Norris Hundley Jr., *Great Thirst*, 213–19.

10. Hoover, *Future Development*.

11. Hundley, *Great Thirst*, 248–52.

12. *Los Angeles Tribune*, 1 July 1917, box 12, Strong Collection, Huntington Library.

13. "From Mountain Top to Ocean in Controlling Floods," manuscript, January 28, 1916, box 12, Strong Collection, Huntington Library.

14. Letter from Harriet Strong to Woodrow Wilson, 21 August 1918, box 14, Strong Collection, Huntington Library; "From Mountain Top to Ocean in Controlling Floods"; copy of a letter Strong sent to senators and representatives in Washington, D.C., regarding HR 15777, 30 January 1919, box 14, Strong Collection; H.R. 15777: *A Bill to Provide for Flood Control on the Colorado River* . . . Feb. 7, 1919, 65th Congress, 3d Session.

15. Letter from Harriet Strong to Woodrow Wilson, 21 August 1918.

16. *Commonwealth* 7 (2 June 1931): 57.

17. Millard Peck, "Reclamation Projects and Their Relation to Agricultural Depression," *Annals of the American Academy of Political and Social Science* 142 (March 1929): 182.

18. Ibid.; Frank P. Willits, "The Futility of Further Development of Irrigation Projects," in Millard Peck, "Reclamation Projects," 186–95.

19. University of California Agricultural Experiment Station, "Economic Problems of California Agriculture," 37. No less an authority than Secretary of Agriculture Arthur Hyde wrote in 1930: "Expansion has been misdirected as well as overstimulated." Arthur M. Hyde, "Report of the Secretary," *Yearbook of Agriculture* (Washington D.C.: Government Printing Office, 1930), 36–37.

20. University of California Agricultural Experiment Station, "Economic Problems of California Agriculture," 39; Samuel N. Dicken, "Dry Farming in the San Joaquin, California," *Economic Geography* 8 (1932): 98.

21. Ibid., 99.

22. Baker, "Changes in Production," 145.

23. Ibid., 146.

24. Lindsay *Gazette*, 25 April 1920.

25. *Progressive Map of Fresno, California: Ownership Atlas of Fresno County and Part of Tulare County* ([San Francisco?]: Progressive Map Service, [1926?]).

26. California Orchard Company, "Letter to Stockholders," 28 March 1924, Frank Honeywell Papers: 1926–31, Bancroft.

27. University of California Agricultural Experiment Station, *Economic Problems of California Agriculture*, 38.

28. "Importance of the Water Situation," in *Introductory Information on the California Water Situation*, Research Department of the California Development Association, Frank Adams Papers, folder 280, WRCA.

29. J. Karl Lee, "Economies of Scale," figure 1, "Changes in Ground Water Levels, 1923–1942, Southern San Joaquin Valley, California."

30. *Yearbook of Agriculture, 1925*, 1390.

31. Herman Steen, "Taking the Premium Out of Capital," *Wallaces' Farmer*, 12 November 1920.

32. Eugene Meyer Jr., managing director of the United States War Finance Corporation, report to President Harding, May 1922, quoted in R. E. Hodges and E. J. Wickson, *Farming in California* (San Francisco: Californians, 1923), 42.

33. "The Progress of the World," *American Review of Reviews* 72 (August 1925): 118–19.

34. Ibid., 121.

35. Letter, George Hecke to Frank Honeywell, editor of the *Pacific Rural Press*, 18 June 1926, in Frank Honeywell Papers: 1926–31, Bancroft Library.

36. Letter, George Hecke to Frank Honeywell, 12 April 1928, Frank Honeywell Papers: 1926–31, Bancroft Library.

37. Willard W. Cochrane, *Development of American Agriculture*, 428–29.

38. R. L. Adams, "The Management of Large Farms," *Agricultural Engineering* 12 (September 1931): 353.

39. Ibid., 354.

40. United States Department of Commerce, Bureau of the Census, *Fifteenth Census of the United States: 1930, Census of Agriculture, Large-Scale Farming*, 1–4.

41. United States Department of Commerce, Bureau of the Census, *Fifteenth Census of the United States: 1930, Census of Agriculture, Large-Scale Farming*, tables 1 and 3.

42. Wendell Berry, "Whose Head Is the Farmer Using? Whose Head Is Using the Farmer?" in Wes Jackson, Wendell Berry, and Bruce Colman, eds., *Meeting the Expectations of the Land* (San Francisco: North Point Press, 1984), 17–30, 24.

43. United States Department of Commerce, Bureau of the Census, *Fifteenth Census of the United States: 1930, Agriculture, California* (Government Printing Office, 1931), 32–34.

44. E. G. Nourse, "Economic Issues of Large-Scale Farming," *Agricultural Engineering* 10 (January 1929): 16; E. G. Nourse, "The Apparent Trend of Recent Economic Changes in Agriculture," *Annals of the American Academy* 144 (May 1930): 46–47.

45. Ibid., 47. My emphasis.

46. United States Department of Commerce, Bureau of the Census, *Fifteenth Census of the United States: 1930, Agriculture, California*, 5, 40–43. Total farms: 135,676. Total acres under cultivation: 30,442,581. One grower advised his colleagues to "select as foremen over your cutting shed and over other departments men who have had previous experience in the handling of peaches, and whom you know to be possessed of a sufficient amount of executive ability to compel the workers to do their work properly." "Handling the Peach Crop," *Associated Grower* 1 (July 1920): 7.

47. Frank Adams, "Third Talk—Demonstration Train," manuscript address (1928), Adams Papers, folder 146, WRCA.

48. Wheeler McMillen, "Engineers, Farmers, and Tomorrow," *Agricultural Engineering* 11 (August 1930): 270.

49. C. C. Teague believed that his holdings allowed him to beat out the small-fry competition in ways that had nothing to do with efficiency: "I am sure that large growers, such as we are, can make a good thing out of the lemon business when the little fellow will starve to death." Letter, C. C. Teague to Waldo A. Hardison, 12 July 1902, C. C. Teague Papers, Bancroft Library.

50. Richard Adams, "Management of Large Farms."

51. Lee, "Economies of Scale," 117.

52. Nourse, "Large-Scale Farming," 15; Gene Logsdon suggests that the move to large scale appealed to farmers eager to fend off the perception of unsophistication: "I suspect that . . . farmers continue to expand their farm empires not out of greed or an insatiable desire for wealth, but because they feel compelled to prove again and again that, by God, they are not inferior to anyone." Gene Logsdon, *At Nature's Pace: Farming and the American Dream* (New York: Random House, 1994), 48.

53. George N. Peek and Hugh S. Johnson, *Equality for Agriculture* (Moline, Ill.: H. W. Harrington, 1922), 5–7, 11, 14–19. Also See David Hamilton, *From New Day to New Deal: American Farm Policy from Hoover to Roosevelt, 1928–1933* (Chapel Hill: University of North Carolina Press, 1991), 5; and Gilbert Fite, *George N. Peek and the Fight for Farm Parity* (Norman: University of Oklahoma Press, 1954) and Fite, *American Agriculture and Farm Policy since 1900*, Publication Number 59 (New York: American Historical Association, 1964), 11. Two bills passed Congress in 1927 and 1928.

54. Calvin Coolidge, "The Farmer and the Nation," speech delivered at Chicago, Illinois, December 1925, in Janet Podell and Steven Anzovin, eds., *Speeches of the American Presidents* (New York: H. W. Wilson, 1988), 439–41.

55. Calvin Coolidge, veto address to the Senate of 25 February 1927, quoted in the *New York Times* (26 February 1927).

56. *New York Times* (27 February 1927).

57. Robert S. Brookings, *Agricultural Corporations: The Conversion of Agriculture into a Prosperous Industry* (Washington, D.C.: Robert S. Brookings, 1928), 13. Printed together with *Socializing the Soulless Corporation: A Sequel to Agricultural Corporations*.

58. Henry Ford, in collaboration with Samuel Crowther, *Today and Tomorrow* (Garden City: Doubleday, Page, and Company, 1926), 211.

59. Ibid., 215–16. Ford gave his blessing to monoculture as the most efficient use of the farmer's time: "The single-crop farmer is through for the year with a month or half a month's work," he said, "during the remainder of the time he watches nature work for him." Ibid., 212, 213.

60. Ibid., 218.

61. Edward Sherwood Mead and Bernhard Ostrolenk, *Harvey Baum: A Study of the Agricultural Revolution* (Philadelphia: University of Pennsylvania Press, 1928), 145.

62. Ibid., 140–41.

63. Bernhard Ostrolenk, *The Surplus Farmer* (New York: Harper and Brothers, 1932), 34. Also see O. W. Willcox, *Reshaping Agriculture* (New York: W. W. Norton, 1934) and Russell C. Engberg, *Industrial Prosperity and the Farmer* (New York: Macmillan, 1928).

64. Willcox, *Reshaping Agriculture*, 78.

65. Ibid., 153.

66. Ibid., 156.

67. Hamilton, *From New Day;* Sally H. Clarke, *Regulation and the Revolution in United States Farm Productivity* (Cambridge: Cambridge University Press, 1994), 74–75, 139.

68. Will Rogers, "Florida versus California: A Debate Held before the Prevaricators' Club of America," *Saturday Evening Post* 198 (29 May 1926): 10.

69. Hal Roach (producer) and George Jeske (director). Stan Laurel and Katherine Grant in *Oranges and Lemons*, 1923.

70. Adolph Zukor (producer) and Norman McLeod (director). W. C. Fields and Kathleen Howard in *It's a Gift*, Paramount Pictures, 1934.

71. McWilliams, *Factories in the Field*, 231–39, 250–51.

72. Simon Lubin, "Can the Radicals Capture the Farms in California?" speech delivered before the Commonwealth Club of California, March 1934, Simon Lubin Papers, Bancroft Library.

73. United States Congress, Senate Subcommittee of the Committee on Education and Labor, *Violations of Free Speech and Rights of Labor*, Hearings Pursuant to Senate Resolution 266, 76th Cong., 2nd Sess., Pt. 48 (Washington D.C.: Government Printing Office, 1940), 17647–651. Senator Elbert Thomas of Utah served as chair of the whole Committee on Education and Labor.

74. "Who Are the Associated Farmers?" *Rural Observer* 1 (September–October 1938). Published by the Simon J. Lubin Society of California, Inc.

75. Ibid.

76. *Rural Observer* 1 (March 1938).

77. McWilliams, *Factories in the Field*, 251.

78. McWilliams was born in 1905 in Steamboat Springs, Colorado. He graduated with an LL.B. from University of Southern California in 1927 and later became a contributing editor at the *Nation*. His first book, *Ambrose Bierce: A Biography* was published in 1929.

79. John Steinbeck, *The Grapes of Wrath* (Viking Press, 1939; reprint, New York: Penguin Books, 1985), 224.

80. Ibid., 224–27.

81. Mead and Ostrolenk, *Harvey Baum*, 149.

82. E. G. Nourse, "Revolution in Farming," 90–91.

Epilogue: Restless Orchard

1. Walter Goldschmidt, *As You Sow* (New York: Harcourt, Brace, and Company, 1947), 203–10, 221–38.

2. L. H. Bailey, quoted in Dwight Sanderson, ed., *Farm Income and Farm Life: A Symposium on the Relation of the Social and Economic Factors in Rural Progress* (Chicago: University of Chicago Press, 1927), 10. Bailey's emphasis.

3. Liberty Hyde Bailey, *The Garden Lover* (New York: Macmillan, 1928), 7.

4. Ibid.

5. E. G. Nourse, "The Apparent Trend of Recent Economic Changes in Agriculture," *The Annals of the American Academy* 44 (May 1930): 46.

6. E. G. Nourse, "The Farmer's Mysterious Malady," unpublished article quoted in Knapp, *Edwin G. Nourse*, 109–10.

7. Illustration by Gib Crockett from the *Washington Evening Star* dated 1947, quoted in Knapp, *Edwin G. Nourse*, plates following 272.

8. Miguel A. Altieri, *Agroecology: The Science of Sustainable Agriculture* (Boulder, Colo.: Westview Press, 1995), 369, 271–73, 214–15; Robert E. Rhoades, "The World's Food Supply at Risk," *National Geographic* 179 (April 1991): 75.

9. Rhoades, "Food Supply," 83.

10. For a discussion of recent innovations in rice breeding that use land races and thus resist the trend to genetic erosion, see Robert E. Evenson and Douglas Gollin, *Genetic Resources, International Organizations, and Rice Varietal Improvement* (New Haven: Economic Growth Center, Yale University, 1994).

11. Jim Hightower, *Hard Tomatoes, Hard Times: The Original Hightower Report—and Other Recent Reports—on Problems and Prospects of American Agriculture* (Cambridge, Mass.: Schenkman, 1978), 5. The internal quotation is from Dr. Earl L. Butz, former dean of the College of Agriculture at Purdue University and U.S. secretary of agriculture in 1955; Kamyar Enshayan, *Dr. Twisted Visits a Farm* (Cedar Falls, Iowa: self-published, 1994), passim.

12. Duncan, *Centrality of Agriculture*.

13. Thoreau said "in Wildness is the preservation of the World." For a critique of wilderness see William Cronon, "The Trouble with Wilderness; or, Getting Back to the Wrong Nature," in William Cronon, ed., *Uncommon Ground: Toward Reinventing Nature* (New York: W. W. Norton and Company, 1995), 89. For an argument in favor of agriculture as the core of a sustainable society see Duncan, *Centrality of Agriculture*.

There are other subjects that bear on the present one, such as the tension between a rising world population and the energy-intensive methods we employ to create our food, as well as the disappearance of farmland through soil erosion and real estate development—all too complicated to be considered here.

14. Raymond Williams, *The Country and the City* (New York: Oxford University Press, 1973), 297.

Bibliography

1. Manuscript Collections

California State Archives
Economic Poison Certificates, 1921–22.
Records of the California Department of Agriculture, 1930–40.
Chevron Corporate Library, San Francisco, California
California Spray-Chemical Company. Papers, 1907–29.
Bancroft Library, University of California at Berkeley
Adams, Frank. Papers. Water Resources Center Archives.
Blowers, R. B. and Sons. Business Records and Papers
California Pear Growers' Association. Minutes, 1916–25.
Clements, George P. Papers, 1920–46.
DiGiorgio, Robert, and Joseph A. DiGiorgio, *The DiGiorgios: From Fruit Merchants to Corporate Innovators*. Interview conducted by Ruth Teiser, 1983. Regional Oral History Office, University of California at Berkeley.
Drobish, Harry Everett. Papers, 1917–54.
Eisen, Gustav. Papers, 1890–1900.
Erdman, Henry Ernest. Papers, 1880–1930.
Federal Writers Project of Oakland. Papers.
Honeywell, Frank. Papers, 1926–31.
Johnson, Hiram. Papers, 1895–1945.
Lambert, Charles F. "Land Speculation and Irrigation in the Sacramento Valley, 1905–1957." Interview conducted by Willa Baum. Berkeley: University of California General Library Regional Cultural History Project, 1957.
Lubin, Simon J. Papers.
Rowell, Chester H. Papers, 1887–1946.
Swett, Frank T. "California Agricultural Cooperatives, Interview with Frank T. Swett." Conducted by Willa Klug Baum. Berkeley: Regional Oral History Project, 1968.
Teague, Charles C. Correspondence and Papers, 1901–50.

University of California Archives, Records of the College of Agriculture
    and the Agricultural Experiment Station.
Weinstock, Harris. Scrapbooks, 1893–1922.
Wickson, Edward James. Papers, 1868–1923.
Woodworth, Charles William. Papers, 1899–1937.
Henry E. Huntington Library, San Marino, California.
    Carlson, Charles. Collection.
    Latta, Frank. Collection and "Sky-Farming" Papers.
    Strong, Harriet Williams Russell. Collection, 1815–1939.
Juda L. Magnes Museum, Berkeley, California.
    Lubin, David. Papers, 1848–1919.
Stanford University, Green Library, Special Collections.
    Irish, John P. Papers.
    Jones, Herbert C. Papers, 1903–54.
Yale University, Sterling Library, Microfilm Collection.
    Roosevelt, Theodore. Papers.

2.   Periodicals

Agricultural Engineering
American Academy of Political and Social Science. *Annals.*
American Economic Review
American Geographical Society. *Bulletin.*
American Review of Reviews
Analyst
Appleton's Journal
Associated Grower
Association of American Geographers. *Annals.*
California—A Journal of Rural Industry
California Bureau of Labor Statistics, *Biennial Report*
California Citrograph
California Citrus Institute. *Annual Report* (San Bernardino).
California Cultivator
California Fruit Growers. *Proceedings.*
California Fruit Growers. *Report.*
California Fruit Growers. *Transactions.*
California Horticulturist and Floral Magazine
California Pear Growers' Association. *Annual Report.*
California Pear Growers' Association. *California Pear Grower.*
California State Agricultural Society. *Transactions.*
California State Commission of Horticulture. *Monthly Bulletin.*
California State Fruit Growers' Convention. *Report.*
California State Horticultural Society. *Proceedings.*
California State Market Director. *Annual Report.*
Commission of Immigration and Housing. *Annual Report.*
Commonwealth Club of California. *Transactions.*

Current History
Economic Geography
Entomological Society of America. *Annals.*
Farmers News
Fortune
Fresno Expositor.(Fresno, California)
Fresno Morning Republican (Fresno, California)
Irrigation Age
Journal of Economic Entomology
Journal of Political Economy
Lindsay Gazette. (Lindsay, California)
Literary Digest
Maywood Colony Advocate
New York Times
Overland Monthly
Pacific Rural Press
Quarterly Journal of Economics
Rural Observer
Sacramento Bee
San Francisco Chronicle
Saturday Evening Post
Scientific American
Simon J. Lubin Society of California. *The Rural Observer.*
State Fruit-Growers' Convention. *Proceedings.*
State Fruit-Growers' Convention. *Transactions.*
United States Department of Agriculture. *Yearbook.*
University of California Agricultural Experiment Station. *Bulletin.*
University of California Agricultural Experiment Station. *Circular.*
Vedalia
Wallace's Farmer
Western Canner and Packer
Western Fruit Jobbers Association. *Proceedings.*
World's Work
Yale Review

3. Rare Pamphlets

Abbreviations

Bancroft   Bancroft Library, University of California at Berkeley
Beinecke   Beinecke Rare Book Library, Yale University
UCB   University of California at Berkeley Library

Adams, Richard L. *Farm Management Notes (for California).* 7th ed. Berkeley: Associated Students Store, 1921. Bancroft.
Baker, P. Y. *The 76 Land and Water Company's Lands.* San Francisco: n.p., [1884?]. Beinecke.

Braun Corporation. *School of Fumigation . . . Held at Pomona, California, August 9–13, 1915*. Los Angeles: Braun Corporation, 1915. Bancroft.

Bremner, O. E. *Destructive Insects and Their Control*. Sacramento: Superintendent of State Printing, 1910. UCB.

Brookings, Robert S. *Agricultural Corporations: The Conversion of Agriculture into a Prosperous Industry*. Printed together with *Socializing the Soulless Corporation: A Sequel to Agricultural Corporations*. Washington, D.C.: Robert S. Brookings, 1928. Yale University Library.

Brown, Seth R. *Report of the Division of State Employment Agencies of the Department of Industrial Relations of the State of California*. Sacramento: State Printing Office, 1928. UCB.

"Buhach" The California Insecticide, pamphlet. N.p., n.d. Bancroft.

Butterfield, Harry M. "Dates of Introduction of Trees and Shrubs to California." Manuscript produced by the Department of Landscape Horticulture, University of California at Davis, 1964. Bancroft.

California Bureau of Labor Statistics. *Private Employment Agency Law*, San Francisco: State Printing Office, 1924.

California Development Association. *Report on Problems of Agricultural Development in California, Prepared by the Department of Research and Information*. San Francisco: California Development Association, 1924. Bancroft.

———. *Survey of the Mexican Labor Problem in California*. [San Francisco]: California Development Association, [1928]. Bancroft.

California Fruit Growers and Shippers Association. Transportation Committee. *Memorial from the Fruit Growers of California to the Transportation Companies, Asking for a Quicker Time to Points in the East, and for Amendment of the Minimum Carload*. [Sacramento?]: n.p., 1901. Bancroft.

California Growers and Shippers Protective League. *Six Cars of Plums and Other Deciduous Fruits from Central and Northern California Points to Chicago and New York City, June 9th to July 1, 1924*. San Francisco: California Growers and Shippers Protective League, 1924. UCB.

California Pear Growers' Association. *More Buyers for More Pears*. [Sacramento]: California Pear Growers' Association, 1923. Bancroft.

[California State Horticultural Society]. *Tabulated Statement Showing Shipments of California Green Deciduous Fruit Made to Eastern Markets, and the Results of Its Sale Therein*. [Sacramento]: n.p., 1902. UCB.

Cobb, N. A. *The California Wheat Industry*, Miscellaneous Publication No. 159. Sydney, New South Wales, Australia: Department of Agriculture, 1901. Bancroft.

Cooper, Ellwood, *Bug vs. Bug: Parasitology*. N.p., 1913. UCB.

Cramp, Kathryn, Louise F. Shields, and Charles A. Thomson. *Study of the Mexican Population in Imperial Valley, California*. New York: Committee on Farm and Cannery Migrants, Council of Women for Home Missions, 1926. Bancroft.

Craw, Alexander. *Destructive Insects, Their Natural Enemies, Remedies and Recommendations*. Sacramento: State Printing Office, 1891. UCB.

Dean Witter and Company. *California Packing Corporation: A Study of Im-*

pressive Progress. San Francisco: Dean Witter and Company, 1950. UCB.

Dowsett, C. F. *A Start in Life: A Journey Across America. Fruit Farming in California*. London: C. F. Dowsett and Company, 1891. Bancroft.

Fresno County, California, Its Offering for Settlement. San Francisco: Pacific Coast Land Bureau, 1886. Beinecke.

Fresno County, California—Where Can Be Found Climate, Soil, and Water, the Only Sure Combination for the Vineyardist ([San Francisco]: Vogel, Hall, Lisenby), 1887. Bancroft.

Gardner, Leo. *The First Thirty Years: The Early History of the Company Now Known as Ortho Division, Chevron Chemical Company*. [San Francisco?]: Chevron Chemical, 1978. Bancroft.

Hecke, George. *The Mediterranean Fruit Fly: A Fruit Enemy Which Must Be Kept Out of California*. Sacramento: State Department of Agriculture, 1929. UCB.

Hoover, Herbert. "The Future Development of the Great Central Valley of California." Printed address delivered before the Sacramento Chamber of Commerce, 27 June 1925. Bancroft.

Hunt, Thomas Forsyth. *How Can a Young Man Become a Farmer*. Paper read before the San Jose Meeting of the Patrons of Husbandry of California, 21 October 1913. Bancroft.

Isaac, John. *Bug vs. Bug: Nature's Method of Controlling Injurious Species*. Sacramento: State Horticultural Commission, 1906. UCB.

J. Walter Thompson Company. "Presentation on California Fresh Bartlett Pears." Manuscript. San Francisco, 1927. Library of the Giannini Foundation for Agricultural Economics, University of California at Berkeley.

Kearney, M. Theodore. "Fresno County, California: The Land of Sunshine, Fruits and Flowers." In *How to Make Money in California*. Fresno: self-published, 1893. Bancroft.

Lubin, Simon J. *Can the Radicals Capture the Farms in California?* Printed address delivered before the Commonwealth Club of California, San Francisco, March 1934. Bancroft.

Paige, Timothy. *Farming That Pays*. San Francisco: n.p., 1891. Beinecke.

Phillips, R. G. *Wholesale Distribution of Fresh Fruits and Vegetables*. Rochester, New York: The Joint Council of the National League of Commission Merchants of the United States, the Western Fruit Jobbers' Association of America, and the International Apple Shippers' Association, 1922. Yale University Library.

Pixley, Frank. "An Idyl of the Tulare Lake Land." In *Secure a 20-Acre Home in the Tulare Colony*. San Francisco: Pacific Coast Land Bureau, 1885. Huntington Library, San Marino, California.

Price List of Trumbull and Beebe, Fruit and Ornamental Trees. San Francisco: Trumbull and Beebe, 1891. Bancroft.

Rohrabacher, C. A. *Fresno County, Her Live Towns, Her Industries, and Her Progressive Men*. Fresno: n.p., 1891. Beinecke.

Sherwin-Williams Company. *Spraying: A Profitable Investment*. Cleveland: Sherwin-Williams Company, [1918?]. UCB.

Swett, Frank T. *The Pear Growers' Progress: Address of Frank T. Swett*. San Francisco: California State Pear Growers' Association, 1920. Bancroft.

Vivian, Thomas J. *The Commercial, Industrial, Agricultural, Transportation, and Other Interests of California*. Washington: Bureau of Statistics, Treasury Department, 1891. Bancroft.

Wickson, Edward James. *Cost of Planting and Care of Trees and Vines to Bearing Age*. Manuscript, Giannini Foundation of Agricultural Economics Library, Berkeley, California.

4.   Maps

Committee on Phylloxera, Vine Pests, and Diseases of the Vine for the State Board of Viticultural Commissioners. *Viticultural Map*. San Francisco: Britton and Rey, Lithographers, 1880.

The Garden of the World. San Francisco: N. C. Carnall and Company, Dealers in Colony lands, [1890?].

Official Map of the County of Solano, California. [San Francisco?]: E. N. Eager, 1909.

Official Map of the County of Yolo, California. San Francisco: P. N. Ashley, 1909.

Progressive Map of Fresno, California: Ownership Atlas of Fresno County and Part of Tulare County. [San Francisco]: Progressive Map Service, [1926?].

Thompson, Thomas H. *Official Atlas Map of Fresno County*. Fresno: Office of the Board of Supervisors, 1891.

5.   Paintings and Films

Hill, Thomas. *Irrigation at Strawberry Farm* (1865). Bancroft Library.

Roach, Hal (producer), George Jeske (director). Stan Laurel and Katherine Grant in *Oranges and Lemons*, 1923.

Zukor, Adolph (producer), Norman McLeod (director). W. C. Fields and Kathleen Howard in *It's a Gift*. Paramount Pictures, 1934.

6.   Interviews

Folsom, Deborah, botanist at the Henry E. Huntington Library, San Marino, California. Interview by author, San Marino, 9 August 1991.

Gardner, Leo, employee of California Spray-Chemical Company between 1923 and 1969. Interview by author, Alameda, California, 23 September 1992 and 8 February 1993 (the second by telephone).

Heine, Dale, professor of agricultural economics, University of California at Davis. Interview by author, 6 May 1991 (by telephone).

Sussex, Ian, professor of plant genetics, University of California at Berkeley. Interview by author, Berkeley, 3 March 1993.

7.   Experiment Station Publications
     and Selected Government Documents

Adams, Frank. "The Agricultural Situation in California." *University of Cali-
fornia Agricultural Experiment Station, Agricultural Extension Service
Circular No. 18* (1928).
Adams, Richard L., and T. R. Kelly. "A Study in Farm Labor." *University of
California Agricultural Experiment Station Circular No. 193* (March 1918).
Baker, Oliver Edwin, ed. *A Graphic Summary of American Agriculture, Based
Largely on the Census of 1920. United States Department of Agriculture Year-
book, 1921.* Washington, D.C.: Government Printing Office, 1921: 407–506.
Brown, Seth R. *Report of the Division of State Employment Agencies of the
Department of Industrial Relations of the State of California.* Sacramento:
State Printing Office, 1928.
Bureau of Labor Statistics. *Second Biennial Report of the Bureau of Labor Sta-
tistics of the State of California for the Years 1885–1886.* Sacramento: Super-
intendent of State Printing, 1887.
California Board of Horticultural Commissioners, *Meeting* (1882?).
California Commission on Immigration and Housing. *Report on Unemploy-
ment to His Excellency, Governor Hiram W. Johnson.* Sacramento: State
Printing Office, December 9, 1914.
California State Board of Horticulture. *Report on the Importation of Parasites
and Predaceous Insects.* Sacramento: State Printing Office, 1892.
California State Commission of Horticulture. *Horticultural Statutes and
County Ordinances of California in Force July 1, 1905.* Sacramento: Super-
intendent of State Printing, 1905.
Colby, George E. "Arsenical Insecticides: Paris Green; Commercial Substi-
tutes; Home-Made Arsenicals." *University of California Agricultural
Experiment Station Bulletin No. 151* (1903).
Dwinell, C. H. "Entomology in the College of Agriculture." *University of
California Agricultural Experiment Station Bulletin No. 16* (1884).
Enos, John Summerfield. *Statement of John Summerfield Enos to the California
State Horticultural Society.* Sacramento: State Printing Office, 1886.
———. Report. *Second Biennial Report of the Bureau of Labor Statistics of the
State of California for the Years 1885–1886.* Sacramento: Superintendent of
State Printing, 1887.
*First Annual Report of the State Market Director of California to the Governor
of California, for the Year Ending December 1, 1916.* Sacramento: State
Printing Office, 1916.
Hecke, George. "Argument Against the Eight Hour Law," in *Amendments to
the Constitution and Proposed Statutes with Arguments Respecting the Same.*
Sacramento: State Printing Office, 1914.
Hedrick, Ulysis Prentiss. "The Fruit Districts of New York," *The Fruit Indus-
try of New York, Department of Agriculture of the State of New York Bulletin
No. 79* (January 1916).
Hilgard, Eugene W. "Repression of the Phylloxera," in *University of California,
College of Agriculture, Report of the Viticultural Work during the Seasons
1883–4 and 1884–5.* Sacramento: State Printing Office, 1886.

————. *University of California, College of Agriculture, Report of the Viti-cultural Work during the Seasons 1885 and 1886* (Sacramento: State Printing Office, 1886.

Hunt, Thomas Forsyth. "Suggestions to the Settler in California." *University of California Agricultural Experiment Station Circular No. 210* (March 1919).

Johnson, Sherman. *Changes in Farming in War and Peace*, Washington, D.C.: United States Bureau of Agricultural Economics, 1946.

Klee, W. G. "The Woolly Aphis and Its Repression." *University of California Agricultural Experiment Station Bulletin No. 55* (1886).

Kraemer, Erich, and H. E. Erdman. "History of Cooperation in the Market-ing of California Fresh Deciduous Fruits." *University of California Agricul-tural Experiment Station Bulletin No. 557* (September 1933).

Miller, G. H. "Operating Costs of a Well-Established New York Apple Or-chard." *United States Department of Agriculture Bulletin No. 130* (21 Au-gust 1914).

Powell, G. Harold. "Fundamental Principles of Co-operation in Agriculture." *University of California Agricultural Experiment Station Circular No. 123* (October 1914).

Rasmussen, M. P. "Some Facts Concerning the Distribution of Fruits and Veg-etables by Wholesalers and Jobbers in Large Terminal Markets." *Cornell University Agricultural Experiment Station, Ithaca, New York Bulletin 494* (October 1929).

Report of the Country Life Commission. United States Senate Document No. 705, 60th Cong., 2nd Sess. Washington, D.C.: Government Printing Office, 1909.

Second Annual Report of the State Market Director, 5–10, Appendix B, "State Market Commission Act of California [Approved 1 June 1917].

Strahorn, A. T. United States Department of Agriculture, Bureau of Soils, Field Operations. *Soil Survey of the Fresno Area*. Washington, D.C.: Gov-ernment Printing Office, 1914.

Taylor, Paul S. "Factors Which Underlie the Infringement of Civil Rights in Industrial Agriculture." In United States Congress, Senate Subcommittee of the Committee on Education and Labor. *Violations of Free Speech and Rights of Labor*. Hearings Pursuant to Senate Resolution 266, Pt. 62, Ex-hibit 9573, 76th Cong., 3rd Sess. Washington D.C.: Government Printing Office, 1940: 22488–922.

Tufts, Warren P. "The Packing of Apples in California." *University of Califor-nia Agricultural Experiment Station Circular No. 178* (October 1917).

United States Congress, House of Representatives, Committee on Immi-gration and Naturalization. *Seasonal Agricultural Laborers from Mexico*. Hearings pursuant to HR 6741. 69th Cong,. 1st Sess. January 28, 1926.

United States Congress, Senate Subcommittee of the Committee on Educa-tion and Labor. *Violations of Free Speech and Rights of Labor*. Hearings Pursuant to Senate Resolution 266, 76th Congress. Washington D.C.: Government Printing Office, 1940.

United States Department of Agriculture, Bureau of Soils. *Reconnaissance Soil Survey of the Lower San Joaquin Valley, California*. Washington, D.C.: Gov-ernment Printing Office, 1918.

United States Department of Agriculture, Weather Bureau. *Climatic Summary of the United States: Section 17—Central California*. Washington, D.C.: Government Printing Office, 1934.

United States Department of Commerce, Bureau of the Census. *Eleventh Census of the United States—Agriculture*. Washington, D.C.: United States Census Office, 1890.

————. *Twelfth Census of the United States: Agriculture, 1900*. Washington, D.C.: Government Printing Office, 1902.

————. *Thirteenth Census of the United States: Agriculture, General Report, 1910*. Washington, D.C.: Government Printing Office, 1913.

————. *Thirteenth Census of the United States, Agriculture*, vol. I, 28. Washington, D.C.: Government Printing Office, 1913.

————. *Fourteenth Census of the United States: Agriculture, 1920*. Washington, D.C.: Government Printing Office, 1922.

————. *Fifteenth Census of the United States: 1930—Agriculture, Volume III, Type of Farm, Part 3—The Western States*. Washington, D.C.: Government Printing Office, 1932.

————. *Fifteenth Census of the United States: 1930, Census of Agriculture, Large-Scale Farming in the United States, 1929*. Washington, D.C.: Government Printing Office, 1930.

University of California Agricultural Experiment Station. *Report of the Work of the Agricultural Experiment Station of the University of California from June 30, 1903, to June 30, 1904*. Sacramento: State Printing, 1904.

————. *Publications of the Agricultural Experiment Station for the Period 1877–1918*. Berkeley: University of California, 1919.

————. "Economic Problems of California Agriculture: A Report to the Governor of California." *University of California Agricultural Experiment Station Bulletin No. 504* (December 1930).

Woodworth, C. W. "Remedies for Insects and Fungi." *University of California Agricultural Experiment Station Bulletin No. 115* (1898).

————. *Cooperative Work in Economic Entomology*. [Washington, D.C.]: United States Department of Agriculture, Office of Experiment Stations, 1903.

————. "Directions for Spraying the Codling-Moth." *University of California Agricultural Experiment Station Bulletin No. 155* (1904).

————. "List of Insecticide Dealers." *University of California Agricultural Experiment Station Circular No. 79* (July 1912).

8.   Articles and Chapters

Almaguer, Tomás. "Racial Domination and Class Conflict in Capitalist Agriculture: The Oxnard Sugar Beet Workers' Strike of 1903." *Labor History* 25 (summer 1984): 325–50.

Bailey, Liberty Hyde. "The Collapse of Freak Farming." *Country Life in America* 4 (1903): 14–16.

————. "Country Living in the Next Generation." *Independent* 85 (March 1916): 336–38.

Baker, Oliver Edwin. "The Increasing Importance of the Physical Conditions in Determining the Utilization of Land for Agriculture and Forest Production in the United States," *Annals of the Association of American Geographers* 11 (1921): 39–46.

———. "The Agriculture of the Great Plains Region," *Annals of the Association of American Geographers* 13 (September 1923): 109–67.

———. "The Agricultural Regions of North America." *Economic Geography* 2 (October 1926): 459–93; 3 (January, July, October 1927): 50–86, 309–39, 447–65; 4 (January, October 1928): 44–73, 399–433; 5 (January 1929): 36–69; and 6 (April 1930): 166–90.

———. "Changes in Production and Consumption of Our Farm Products and the Trend in Population: Do We Need More Farm Land?" *Annals of the American Academy* 149 (May 1930): 97–234.

Baker, Oliver Edwin, and H. M. Strong. "Arable Lands in the United States," *United States Department of Agriculture Yearbook 1918*, 433–41. Washington, D.C.: Government Printing Office, 1919.

Ball, E. D. "Is Arsenical Spraying Killing Our Fruit Trees?" *Journal of Economic Entomology* 2 (April 1909): 142–48.

Ball, E. D., E. G. Titus, and J. E. Greaves. "The Season's Work on Arsenical Poisoning of Fruit Trees." *Journal of Economic Entomology* 3 (April 1910): 187–97.

Berry, Wendell. "Whose Head Is the Farmer Using? Whose Head Is Using the Farmer?" In *Meeting the Expectations of the Land*, edited by Wes Jackson, Wendell Berry, and Bruce Colman. San Francisco: North Point Press, 1984.

Bloom, Khaled. "Pioneer Land Speculation in California's San Joaquin Valley." *Agricultural History* 57 (July 1983): 297–307.

Bogue, Margaret Beattie. "The Lake and the Fruit: The Making of Three Farm-Type Areas." *Agricultural History* 59 (October 1985): 493–522.

Bordwell, Georgia Graves. "Who Says White Folks Won't Work: Substituting American Families for Asiatics in California Orchards." *Sunset* 45 (December 1920): 28–30, 101.

Colby, Charles C. "The California Raisin Industry—a Study in Geographic Interpretation." *Annals of the Association of American Geographers*, 14 (June 1924): 50–107.

Cox, Lawanda F. "The American Agricultural Wage Earner, 1865–1900: The Emergence of a Modern Labor Problem." *Agricultural History* 22 (April 1948): 100–103.

Cross, Ira B. "Cooperation in Agriculture." *American Economic Review* 1 (September 1911): 535–44.

Danbom, David B. "Romantic Agrarianism in Twentieth-Century America." *Agricultural History* 65 (fall 1991): 1–12.

Erdman, H. E. "The Development and Significance of California Cooperatives, 1900–1915." *Agricultural History* 32 (1958): 179–84.

Fite, Gilbert C. *American Agriculture and Farm Policy since 1900*. Publication Number 59. Service Center for Teachers of History. New York: American Historical Association, 1964.

FitzSimmons, Margaret. "The New Industrial Agriculture: The Regional

Integration of Specialty Crop Production." *Economic Geography* 62 (October 1986): 334–53.

Francisco, Don. "The Advertising of Agricultural Specialties." *American Co-operation: A Collection of Papers and Discussions Comprising the Fourth Summer Session of the American Institute of Cooperation at the University of California*. Vol. 2. Washington, D.C.: American Institute of Cooperation, 1928.

Friday, David. "The Course of Agricultural Income during the Last Twenty-Five Years." *American Economic Review* 13 (March 1923), supplement: 146–58.

Gates, Paul Wallace. "The Homestead Law in an Incongruous Land System." *American Historical Review* 41 (1935–36) 652–81.

———. "Public Land Disposal in California." *Agricultural History* 49 (January 1975): 158–78.

———. "An Overview of American Land Policy." In *American Law and the Constitutional Order: Historical Perspectives*, edited by Lawrence M. Friedman and Harry N. Scheiber. Cambridge: Harvard University Press, 1978.

George, Henry. "Our Land and Land Policy." Vol. 9 of *Writings of Henry George*. New York: Doubleday and McClure, 1901.

González, Gilbert G. "Company Unions, the Mexican Consulate, and the Imperial Valley Agricultural Strikes, 1928–1934 *Western Historical Quarterly* 27 (spring 1996): 53–73.

Gregor, Howard F. "The Local-Supply Agriculture of California." *Association of American Geographers* 47 (September 1957): 267–76.

Gregson, Mary Eschelbach. "Specialization in Late-Nineteenth-Century Midwestern Agriculture: Missouri as a Test Case." *Agricultural History* 67 (winter 1993): 16–36.

Hackel, Otto. "Summary of the Geology of the Great Valley." In *Geology of Northern California*, edited by Edgar H. Bailey. San Francisco: California State Division of Mines and Geology, 1966.

Haedden, William. "Arsenical Poisoning of Fruit Trees." *Journal of Economic Entomology* 2 (June 1909): 239–45.

Hecke, G. H. "The Pacific Coast Labor Question from the Standpoint of a Horticulturist," in *Official Report of the Thirty-Third Fruit-Growers' Convention*. Sacramento: State Printing Office, 1908.

Jenny, Hans, and collaborators, "Exploring the Soils of California," in Claude B. Hutchinson, ed., *California Agriculture*, Berkeley and Los Angeles: University of California Press, 1946.

Jones, Lamar B. "Labor and Management in California Agriculture, 1864–1964." *Labor History* 11 (winter 1970): 23–40.

Kirkendall, Richard S. "L. C. Gray and the Supply of Agricultural Land." *Agricultural History* 37 (October 1963): 206–14.

Moses, H. Vincent. "'The Orange-Grower Is Not a Farmer': G. Harold Powell, Riverside Orchardists, and the Coming of Industrial Agriculture, 1893–1930." *California History* 74 (spring 1995): 22–37.

Nash, Gerald D. "Henry George Reexamined: William S. Chapman's Views on Land Speculation in Nineteenth Century California." *Agricultural History* 33 (1959): 133–37.

Nourse, E. G. "The Aim and Scope of Agricultural Economics." In *Agricultural Economics*, edited by E. G. Nourse. Chicago: University of Chicago Press, 1916.

———. "The Revolution in Farming." *Yale Review*, n.s., 8 (October 1918): 90–105.

———. "The Place of Agriculture in Modern Industrial Society." *Journal of Political Economy* 27 (July 1919): 561–77.

———. "The Economic Philosophy of Cooperation." *American Economic Review* 12 (December 1922): 577–97.

———. "The Apparent Trend of Recent Economic Changes in Agriculture." *Annals of the American Academy* 149 (May 1930): 45–50.

Orsi, Richard J. "*The Octopus* Reconsidered: The Southern Pacific and Agricultural Modernization in California, 1865–1915." *California Historical Quarterly* 54 (fall 1975): 197–220.

Parsons, James J. "The Uniqueness of California." *American Quarterly* 7 (spring 1955): 45–55.

Paul, Rodman Wilson. "The Great California Grain War: The Grangers Challenge the Wheat King." *Pacific Historical Review* 27 (November 1958): 331–49.

———. "The Wheat Trade between California and the United Kingdom." *Mississippi Valley Historical Review* 45 (December 1958): 391–412.

———. "The Beginnings of Agriculture in California: Innovation vs. Continuity." *California Historical Quarterly* 52 (1973): 16–27.

Peck, Millard. "Reclamation Projects and Their Relation to Agricultural Depression." *Annals of the American Academy of Political and Social Science* 142 (March 1929): 177–85.

Pisani, Donald. "Land Monopoly in Nineteenth-Century California," *Agricultural History* 65 (fall 1991): 15–37.

Powell, Fred Wilbur. "Co-operative Marketing of California Fresh Fruits." *Quarterly Journal of Economics* 25 (February 1910): 392–418.

Rhoades, Robert E. "The World's Food Supply at Risk." *National Geographic* 179 (April 1991): 74–105.

Rice, T. D., and J. A. Warren. "Relation of Agricultural Industries to Natural Environment in the Great Plains." In *Source Book for the Economic Geography of North America*, edited by Charles C. Colby. 3rd ed. Chicago: University of Chicago Press, 1926.

Rome, Adam Ward. "American Farmers as Entrepreneurs, 1870–1900." *Agricultural History* 56 (January 1982): 37–49.

Roosevelt, Theodore. "Seventh Annual Message, December 3, 1907." Vol. 3 (1905–66) of *The State of the Union Messages of the Presidents, 1790–1966*, edited by Fred L. Israel. New York: Chelsea House Publishers, 1967.

Rothstein, Morton. "A British Firm on the American West Coast, 1869–1914." *Business History Review* 37 (winter 1963): 392–415.

———. "The American West and Foreign Markets, 1850–1900." In *The Frontier in American Development*, edited by David M. Ellis. Ithaca: Cornell University Press, 1969.

Sandberg, Gary R., and Ingrid K. Allen. "A Proposed Arsenic Cycle in an Agronomic Ecosystem." In *Arsenical Pesticides*, edited by E. A. Woolson.

American Chemical Society Symposium Series 7. Washington, D.C.: American Chemical Society, 1975.

Seftel, Howard. "Government Regulation and the Rise of the California Fruit Industry: The Entrepreneurial Attack on Fruit Pests, 1880–1920." *Business History Review* 59 (autumn 1985): 369–402.

Smith, Alice Prescott. "The Battalion of Life: Our Woman's Land Army and Its Work in the West." *Sunset* 41 (November 1918): 30–33.

Smith, Ralph E., Harry S. Smith, Henry J. Quayle, and E. O. Essig. "Protecting Plants from Their Enemies." In *California Agriculture*, edited by Claude B. Hutchinson. Berkeley and Los Angeles: University of California Press, 1946.

Stoll, Steven. "Insects and Institutions: University Science and the Fruit Business in California." *Agricultural History* 69 (spring 1995): 216–39.

Taylor, Paul S. "Foundations of California Rural Society." *California Historical Society Quarterly* 24 (1945): 193–228.

Taylor, Paul, and Tom Vasey. "Historical Background of California Farm Labor." *Rural Sociology* I (September 1936): 281–95.

Thickens, Virginia E. "Pioneer Agricultural Colonies of Fresno County." *California Historical Society Quarterly* 25 (1946): 169–77.

Thompson, John G. "The Nature of Demand for Agricultural Products and Some Important Consequences." *Journal of Political Economy* 24 (February 1916): 158–82.

Vandecaveye, S. C., G. M. Horner, and C. M. Keaton. "Unproductiveness of Certain Orchard Soils as Related to Lead Arsenate Spray Accumulations." *Soil Science* 42 (1936): 203–13.

Vaught, David. " 'An Orchardist's Point of View': Harvest Labor Relations on a California Almond Ranch, 1892–1921." *Agricultural History* 69 (fall 1995): 563–91.

Weiman, David. "Staple Crops and Slave Plantations: Alternative Perspectives on Regional Development in the Antebellum Cotton South." In *Agriculture and National Development: Views on the Nineteenth Century*, edited by Lou Ferleger. The Henry A. Wallace Series on Agricultural History and Rural Studies. Ames: Iowa State University Press, 1990.

Woodworth, C. W. "The Insecticide Industries in California." *Journal of Economic Entomology* 4 (August 1912): 358.

9. Books and Dissertations

Adams, Edward F. *The Modern Farmer in His Business Relations*. San Francisco: N. J. Stone, 1899.

Adams, R. L. *Farm Management*. [Berkeley]: n.p., [1918?].

Almaguer, Tomás. *Racial Fault Lines: The Historical Origins of White Supremacy in California*. Berkeley and Los Angeles: University of California Press, 1994.

Altieri, Miguel A. *Agroecology: The Science of Sustainable Agriculture*. Boulder, Colorado: Westview Press, 1995.

Bailey, Liberty Hyde. *The Horticulturist's Rule Book*. New York: Macmillan, 1899.

———. *The Outlook to Nature*. New York: Macmillan, 1905.

———. *The Country-Life Movement in the United States*. 1911. Reprint, New York: Macmillan, 1915.

———. *The Holy Earth*. Ithaca, New York: Comstock Publishing, 1919.

———. *The Harvest of the Year to the Tiller of the Soil*. New York: Macmillan, 1927.

———. *The Garden Lover*. New York: Macmillan, 1928.

Baker, Oliver Edwin, ed. *Atlas of American Agriculture*. Washington, D.C.: Government Printing Office, 1918.

Balderrama, Francisco E., and Raymond Rodríguez. *Decade of Betrayal: Mexican Repatriation in the 1930s*. Albuquerque: University of New Mexico Press, 1995.

Bancroft, Hubert Howe. *History of California*. 7 vols. San Francisco: History Company, 1890.

Barger, Harold, and Hans H. Landsberg. *American Agriculture, 1899–1939: A Study of Output, Employment and Productivity*. New York: National Bureau of Economic Research, 1942.

Barron, Hal S. *Those Who Stayed Behind: Rural Society in Nineteenth Century New England*. Cambridge: Cambridge University Press, 1984.

Barth, Gunther Paul. *Bitter Strength: A History of the Chinese in the United States, 1850–1870*. Cambridge: Harvard University Press, 1964.

Beard, Charles A., and Mary R. Beard. Vol. 2 of *The Rise of American Civilization*. New York: Macmillan, 1927.

Bennett, H. C. *Chinese Labor: A Lecture, Delivered before the San Francisco Mechanics' Institute*. San Francisco: 1870.

Berry, Wendell. *The Unsettling of America: Culture and Agriculture*. San Francisco: Sierra Club Books, 1977.

Berry, Wendell, and Bruce Colman, eds. *Meeting the Expectations of the Land*. San Francisco: North Point Press, 1984.

Bidwell, Percy, and John Falconer. *History of Agriculture in the Northern United States*. Washington, D.C.: Carnegie Institution, 1925.

Bogue, Allan G. *From Prairie to Corn Belt: Farming on the Illinois and Iowa Prairies in the Nineteenth Century*. Chicago: University of Chicago Press, 1963; reprint, Chicago: Quadrangle Paperback, 1968.

Brewer, William H. *Up and Down California in 1860–1864*. Edited by Francis P. Farquhar. Berkeley and Los Angeles: University of California Press, 1966; reprint, 1974.

Bryant, Edwin. *What I Saw in California*. 1848. Reprint, Lincoln: University of Nebraska Press, 1985.

Buechel, Frederick Anthony. *The Commerce of Agriculture: A Survey of Agricultural Resources*. New York: John Wiley and Sons, 1926.

Butterfield, Kenyon L. *Chapters in Rural Progress*. Chicago: University of Chicago Press, 1907.

California Growers and Shippers Protective League. *Refrigeration Test Trip of Six Cars of Plums and Other Deciduous Fruits from Central and Northern*

California Points to Chicago and New York City, June 9th to July 1, 1924. San Francisco: California Growers and Shippers Protective League, 1924.

Campbell, Macy. *Rural Life at the Crossroads.* Boston: Ginn and Company, 1927.

Capper, Arthur. *The Agriculture Bloc.* New York: Harcourt, Brace, and Company: 1922.

Caughey, John, and LaRee Caughey, eds. *California Heritage: An Anthology of History and Literature.* Rev. ed. Itasca, Ill.: F. E. Peacock, Publishers, 1971.

————. *Los Angeles: Biography of a City.* Berkeley and Los Angeles: University of California Press, 1976.

Chambers, Clark A. *California Farm Organizations: A Historical Study of the Grange, the Farm Bureau, and the Associated Farmers, 1929–1941.* Berkeley and Los Angeles: University of California Press, 1952.

Chan, Sucheng. *This Bitter-Sweet Soil: The Chinese in California Agriculture, 1860–1910.* Berkeley and Los Angeles: University of California Press, 1986.

Chandler, Alfred D., Jr. *The Visible Hand: The Managerial Revolution in American Business.* Cambridge: Harvard University Press, 1977.

————, with Takashi Hikino. *Scale and Scope: The Dynamics of Industrial Capitalism.* Cambridge: Belknap Press, Harvard University, 1990.

Chipman, N. P. *Annual Report of the Committee on the Industrial Resources of the State, Annual Report of the California State Board of Trade* [Sacramento, 1892].

————. *Report upon the Fruit Industry of California.* San Francisco: State Board of Trade of California, 1889.

Clarke, Sally H. *Regulation and the Revolution in United States Farm Productivity.* Cambridge: Cambridge University Press, 1994.

Cleland, Robert Glass. *California in Our Time: 1900–1940.* New York: Knopf, 1947.

Cleland, Robert Glass, and Osgood Hardy. *March of Industry.* Los Angeles: Powell Publishing, 1929.

Clough, Charles W., and William B. Secrest. *Fresno County—the Pioneer Years, from the Beginnings to 1900.* Fresno: Panorama West Books, 1984.

Cochrane, Willard W. *The Development of American Agriculture: A Historical Analysis.* 2nd ed. Minneapolis: University of Minnesota Press, 1993.

Colby, Charles C., ed. *Source Book for the Economic Geography of North America.* 3rd ed. Chicago: University of Chicago Press, 1926.

Cook, Sherbourn F. *The Population of California Indians, 1769–1970.* Berkeley and Los Angeles: University of California Press, 1976.

Crampton, Beecher. *Grasses in California.* California Natural History Guides, no. 33. Berkeley and Los Angeles: University of California Press, 1974.

Crissey, Forrest. *Where Opportunity Knocks Twice.* 1910. Reprint, Chicago: Reilly and Britton, 1914.

Cronau, Rudolf. *Our Wasteful Nation: The Story of American Prodigality and the Abuse of Our Natural Resources.* New York: M. Kennerley, 1908.

Cronon, William. *Nature's Metropolis: Chicago and the Great West.* New York: W. W. Norton and Company, 1991.

————. ed., *Uncommon Ground: Toward Reinventing Nature*. New York: W. W. Norton and Company, 1995.

Cross, Ira. *Financing an Empire: History of Banking in California*. 4 vols. Chicago: S. J. Clarke, 1927.

Cumberland, William W. *Cooperative Marketing: Its Advantages as Exemplified in the California Fruit Growers Exchange*. Princeton, New Jersey: Princeton University Press, 1917.

Daggett, Stuart. *Chapters on the History of the Southern Pacific*. New York: Ronald Press, 1922.

Dana, Samuel Trask, and Myron Krueger. *California Lands: Ownership, Use, and Management*. Washington, D.C.: American Forestry Association, 1958.

Danbom, David. *The Resisted Revolution: Urban America and the Industrialization of Agriculture, 1900–1930*. Ames: Iowa State University Press, 1979.

————. *Born in the Country: A History of Rural America*. Baltimore: Johns Hopkins University Press, 1995.

Danhof, Clarence H. *Change in Agriculture: The Northern States, 1820–1870*. Cambridge, Massachusetts: Harvard University Press, 1969.

Daniel, Cletus. *Bitter Harvest: A History of California Farmworkers, 1870–1941*. Ithaca: Cornell University Press, 1981.

Davis, J. H., and R. A. Goldberg. *A Concept of Agribusiness*. Boston: Alpine Press for Harvard University, 1957.

Davis, Mike. *City of Quartz: Excavating the Future in Los Angeles*. New York: Vintage Books, 1992.

Davis, Winfield J. *An Illustrated History of Sacramento County California*. Chicago: Lewis Publishing, 1890.

Deverell, William. *Railroad Crossing: Californians and the Railroad, 1850–1910*. Berkeley and Los Angeles: University of California Press, 1994.

Dorf, Philip. *Liberty Hyde Bailey: An Informal Biography*. Ithaca, New York: Cornell University Press, 1956.

Drake, H. S. *Lindsay-Strathmore Land Development Directory*. Lindsay, Calif.: Hall and Burr, 1929.

Duncan, Colin A. M. *The Centrality of Agriculture: Between Humankind and the Rest of Nature*. Montreal and Kingston: McGill-Queen's University Press, 1996.

Dunlap, Thomas R. *DDT: Scientists, Citizens and Public Policy*. Princeton, N.J.: Princeton University Press, 1982.

Eisen, Gustav. *The Raisin Industry: A Practical Treatise on the Raisin Grapes, Their History, Culture and Curing*. San Francisco: H. S. Crocker and Company, 1890.

Elton, Charles S. *The Ecology of Invasions by Animals and Plants*. New York: John Wiley and Sons, 1958; reprint, 1977.

Ely, Richard T., and George S. Wehrwein. *Land Economics*. 1940. Reprint, Madison: University of Wisconsin Press, 1964.

Enfield, R. R. *The Agricultural Crisis, 1920–1923*. London: Longmans, Green and Company, 1924.

Engberg, Russell C. *Industrial Prosperity and the Farmer*. New York: Macmillan, 1928.

Enshayan, Kamyar. *Dr. Twisted Visits a Farm*. Cedar Falls, Iowa: self-published, 1994.

Erdman, Henry E. *American Produce Markets*. Boston: D. C. Heath and Company, 1928.

Essig, E. O. *Injurious and Beneficial Insects of California*. [Sacramento: State Commission of Horticulture, 1913?].

Evenson Robert E., and Douglas Gollin. *Genetic Resources, International Organizations, and Rice Varietal Improvement*. New Haven: Economic Growth Center, Yale University, 1994.

Faragher, John Mack. *Women and Men on the Overland Trail*. New Haven: Yale University Press, 1979.

————. *Sugar Creek: Life on the Illinois Prairie*. New Haven: Yale University Press, 1986.

Farnsworth, R. W. C. *A Southern California Paradise (in the Suburbs of Los Angeles)*. Pasadena: R. W. C. Farnsworth, 1883.

Federal Writers Project of the Works Progress Administration. *California: A Guide to the Golden State*. New York: Hastings House, 1939; reprint, 1945.

Fisher, Lloyd. *The Harvest Labor Market in California*. Cambridge: Harvard University Press, 1953.

Fite, Gilbert C. *George N. Peek and the Fight for Farm Parity*. Norman: University of Oklahoma Press, 1954.

————. *The Farmers' Frontier, 1865–1900*. New York: Holt, Rinehart, and Winston, 1966.

Fitzgerald, Deborah. *The Business of Breeding: Hybrid Corn in Illinois, 1890–1940*. Ithaca: Cornell University Press, 1990.

Flint, Mary Louise, and Robert van den Bosch. *Introduction to Integrated Pest Management*. New York: Plenum, 1981.

Ford, Henry, in collaboration with Samuel Crowther. *Today and Tomorrow*. Garden City: Doubleday, Page, and Company, 1926.

Fuller, Varden. "The Supply of Agricultural Labor as a Factor in the Evolution of Farm Organization in California." United States Congress, Senate Committee on Education and Labor. Hearings Pursuant to Senate Resolution 266, Exhibit 8762-A. 76th Cong., 3rd Sess. 1940: 19777–898.

Gates, Paul W. *The Farmer's Age: Agriculture 1815–1860*. New York: Holt, Rinehart, and Winston, 1960.

Gates, Paul W., ed. *California Ranchos and Farms, 1846–1862, Including the Letters of John Quincy Adams Warren*. Madison: State Historical Society of Wisconsin, 1967.

Gifford, Edward W., and Gwendoline Harris Block, eds. *California Indian Nights: Stories of the Creation of the World* . . . . Arthur H. Clark, 1930. Reprint, Lincoln: University of Nebraska Press, 1990.

Goldschmidt, Walter. *As You Sow*. New York: Harcourt, Brace, and Company, 1947.

González, Gilbert. *Labor and Community: Mexican Citrus Worker Villages in a Southern California County*. Urbana and Chicago: University of Illinois Press, 1994.

Goodwyn, Lawrence. *Democratic Promise: The Populist Moment in American History*. New York: Oxford University Press, 1976.

————. *The Populist Moment: A Short History of the Agrarian Revolt in America*. New York: Oxford University Press, 1978.

Gregory, James. *American Exodus: The Dust Bowl Migration and Okie Culture in California*. New York: Oxford University Press, 1989.

Guenzil, W. D., ed. *Pesticides in Soil and Water*. Madison, Wisconsin: Soil Science Society of America, 1974.

Guerin-Gonzales, Camille. *Mexican Workers and American Dreams: Immigration, Repatriation, and California Farm Labor, 1900–1939*. New Brunswick, N. J.: Rutgers University Press, 1994.

Gutman, Herbert G. "Social and Economic Structure and Depression in American Labor in 1873 and 1874." Ph.D. diss., University of Wisconsin, 1959.

Haber, Samuel. *Efficiency and Uplift: Scientific Management in the Progressive Era, 1890–1920*. Chicago: University of Chicago Press, 1964.

Hamilton, David. *From New Day to New Deal: American Farm Policy from Hoover to Roosevelt, 1928–1933*. Chapel Hill: University of North Carolina Press, 1991.

Hargreaves, Mary W. M. *Dry Farming in the Northern Great Plains: Years of Readjustment, 1920–1990*. Lawrence: University Press of Kansas, 1993.

Hayter, Earl W. *The Troubled Farmer, 1850–1900: Rural Adjustment to Industrialism*. De Kalb: Northern Illinois University Press, 1968.

Hedrick, Ulysis Prentiss. *A History of Agriculture in the State of New York*. New York: New York State Historical Society, 1933.

Heilbroner, Robert. *The Worldly Philosophers: The Lives, Times and Ideas of the Great Economic Thinkers*. New York: Simon and Schuster, 1953; reprint, 1986.

Heizer, Robert F., and Albert B. Elsasser. *The Natural World of the California Indians*. Berkeley and Los Angeles: University of California Press, 1980.

Hicks, John D. *The Populist Revolt: A History of the Farmers' Alliance and the People's Party*. Minneapolis: University of Minnesota Press, 1931.

Hightower, Jim. *Hard Tomatoes, Hard Times: The Original Hightower Report—and Other Recent Reports—on Problems and Prospects of American Agriculture*. Cambridge, Mass.: Schenkman Publishing, 1978.

Hilgard, Eugene W. *Address on Progressive Agriculture and Industrial Education*. Jackson, Miss.: Clarion Books, 1873.

————. *Report on the Physical and Agricultural Features of the State of California, with a Discussion of the Present and Future Cotton Production in the State*. San Francisco: Pacific Rural Press, 1884.

————. *Soils: Their Formation, Composition and Relations to Climate and Plant Growth*. New York: Macmillan, 1906.

Hilgard, E. W., and W. J. V. Osterhout. *Agriculture for Schools of the Pacific Slope*. New York: Macmillan, 1910.

Hilkert, Richard, and Oscar Lewis. *Breadbasket of the World: California's Great Wheat Growing Era: 1860–1890*. San Francisco: Book Club of California, 1984.

Hill, James J. *Address Delivered by Mr. James J. Hill before the Farmers' National Congress, Madison, Wisconsin, September 24, 1908*. Published separately, but collected in one volume by the Huntington Library, San Marino, California.

Hine, Robert. *California's Utopian Colonies.* Berkeley and Los Angeles: University of California Press, 1954; reprint, New York: W. W. Norton and Company, 1969.

Hodges, R. E., and E. J. Wickson. *Farming in California.* San Francisco: Californians, 1923.

Hofstadter, Richard. *The Age of Reform: From Bryan to F.D.R.* New York: Vintage Books, 1955.

Hornbeck, David, and Phillip Kane, eds. David L. Fuller, cartography. *California Patterns: A Geographical and Historical Atlas.* Mountain View, Calif.: Mayfield Publishing, 1983.

Hundley, Norris, Jr. *Water and the West: The Colorado River Compact and the Politics of Water in the American West.* Berkeley and Los Angeles: University of California Press, 1975.

———. *The Great Thirst: Californians and Water, 1770s-1990s.* Berkeley and Los Angeles: University of California Press, 1992.

Hurtado, Albert L. *Indian Survival on the California Frontier.* New Haven: Yale University Press, 1988.

Hutchinson, Claude B., ed. *California Agriculture,* Berkeley and Los Angeles: University of California Press, 1946.

Ise, John. *Sod and Stubble: The Story of a Kansas Homestead.* Lincoln: University of Nebraska Press, 1936.

Jelinek, Lawrence J. *Harvest Empire: A History of California Agriculture.* San Francisco: Boyd and Fraser, 1979.

Kahrl, William J., ed. *The California Water Atlas.* Governor's Office of Planning and Research, 1979.

Kelley, Robert. *Battling the Inland Sea: American Political Culture, Public Policy, and the Sacramento Valley, 1850–1986* Berkeley and Los Angeles: University of California Press, 1989.

Kellogg, George. *Private Telegraph Code of George Kellogg, Grower and Shipper of Choice Mountain Fruit, New Castle, Placer County, California.* Sacramento: H. S. Crocker, 1893.

Knapp, Joseph G. *Edwin G. Nourse—Economist for the People.* Danville, Ill.: the Interstate Printers and Publishers, 1979.

Kropotkin, P. *Fields, Factories and Workshops.* New York and London: Benjamin Blom, 1913.

Krugman, Paul. *Geography and Trade.* Leuven, Belgium: Leuven University Press; reprint, Cambridge: MIT Press, 1991.

Lamb, Ruth deForest. *American Chamber of Horrors: The Truth About Food and Drugs.* New York: Farrar and Rinehart, 1936.

Lavender, David. *California: Land of New Beginnings.* New York: Harper and Row, 1972; reprint, Lincoln: University of Nebraska Press, 1987.

Lee, J. Karl. "Economies of Scale of Farming in the Southern San Joaquin Valley, California." Manuscript, United States Department of Agriculture, Bureau of Agricultural Economics, Berkeley, April 1946.

Liebman, Ellen. *California Farmland: A History of Large Agricultural Landholdings.* Totowa, N.J.: Rowman and Allanheld, 1983.

Lloyd, John William. *Cooperative and Other Methods of Marketing California*

Horticultural Products. University of Illinois Studies in the Social Sciences, no. 8. Urbana, 1919.

Logsdon, Gene. *At Nature's Pace: Farming and the American Dream*. New York: Random House, 1994.

Maass, Arthur, and Raymond L. Anderson. *. . . And the Desert Shall Rejoice: Conflict, Growth, and Justice in Arid Environments*. Cambridge: MIT Press, 1978.

MacCurdy, Rahno Mabel. *The History of the California Fruit Growers Exchange*. Los Angeles: self-published, 1927.

Majka, Linda C., and Theo J. Majka. *Farm Workers, Agribusiness, and the State*. Philadelphia: Temple University Press, 1982.

Malin, James C. *The Grasslands of North America: Prolegomena to Its History*. Ann Arbor, Mich.: James C. Malin, 1948.

Malone, Michael, and Richard Etulain. *The American West: A Twentieth-Century History*. Lincoln: University of Nebraska Press, 1989.

Marx, Leo. *The Machine in the Garden: Technology and the Pastoral Ideal in America*. New York: Oxford University Press, 1964.

Maskew, Frederick M. *A Sketch of the Origin and Evolution of Quarantine Regulations*. Sacramento: California State Association of County Horticultural Commissioners, 1925.

Mason, A. Freeman. *Spraying, Dusting and Fumigating of Plants: A Popular Handbook on Crop Protection*. New York: Macmillan, 1929.

Mather, Robin. *A Garden of Unearthly Delights: Bioengineering and the Future of Food*. New York: Penguin, 1996.

Maynard, Samuel T. *Successful Fruit Culture: A Practical Guide to the Cultivation and Propagation of Fruits*. New York: Orange and Judd, 1905; reprint, 1913.

McClelland, Gordon T., and Jay T. Last. *California Orange Box Labels: An Illustrated History*. Beverly Hills, Calif.: Hillcrest Press, 1985.

McMillen, Wheeler. *Too Many Farmers: The Story of What Is Here and Ahead in Agriculture*. New York: William Morrow and Company, 1929.

McPhee, John. *Oranges*. New York: Farrar, Straus, Giroux, 1966.

McWilliams, Carey. *Factories in the Field: The Story of Migratory Farm Labor in California*. Boston: Little, Brown, and Company, 1939.

———. *Southern California Country*. New York: Duell, Sloan, and Pearce, 1946.

———. *California: The Great Exception*. New York: Current Books, 1949.

Mead, Edward Sherwood, and Bernhard Ostrolenk. *Harvey Baum: A Study of the Agricultural Revolution*. Philadelphia: University of Pennsylvania Press, 1928.

Miller, Crane S., and Hyslop, Richard S. *California: The Geography of Diversity*. Palo Alto: Mayfield Publishing, 1983.

Mitchell, Don. *Lie of the Land: Migrant Workers and the California Landscape*. Minneapolis: University of Minnesota Press, 1996.

Moses, Herman Vincent. "The Flying Wedge of Cooperation: G. Harold Powell, California Orange Growers, and the Corporate Reconstruction of American Agriculture, 1904–1922." Ph.D. diss., University of California at Riverside, 1994.

Muir, John. *Rambles of a Botanist Among the Plants and Climates of California.* Los Angeles: Dawson's Book Shop, 1974.

Needham, James G., Stuart W. Frost, and Beatrice H. Tothill. *Leaf-Mining Insects.* Baltimore: Williams and Wilkins, 1928.

Neth, Mary. *Preserving the Family Farm: Women, Community, and the Foundations of Agribusiness in the Midwest, 1900–1940.* Baltimore: Johns Hopkins University Press, 1995.

Nordhoff, Charles. *California for Health, Pleasure, and Residence.* 1873. Reprint, Ten Speed Press, 1973.

Nourse, Edwin G., ed. *Agricultural Economics.* Chicago: University of Chicago Press, 1916.

———. *The Legal Status of Agricultural Cooperation.* New York: Macmillan, 1927.

———. *America's Capacity to Produce.* Washington, D.C.: Brookings Institution, 1934.

———. *Marketing Agreements under the Agricultural Adjustment Act.* Washington, D.C.: Brookings Institution, 1935.

Ostrolenk, Bernhard. *The Surplus Farmer.* New York: Harper and Brothers, 1932.

Parker, Carleton H. *The Casual Laborer and Other Essays.* Edited by Cornelia Stratton Parker. New York: Harcourt, Brace, and Howe, 1920.

Paul, Rodman W. *The Far West and the Great Plains in Transition, 1859–1900.* New York: Harper and Row, 1988.

Peek, George N., and Hugh S. Johnson. *Equality for Agriculture.* Moline, Ill.: H. W. Harrington, 1922.

Pisani, Donald J. *From Family Farm to Agribusiness: The Irrigation Crusade in California and the West, 1850–1931.* Berkeley and Los Angeles: University of California Press, 1984.

———. *To Reclaim a Divided West: Water, Law, and Public Policy, 1848–1902.* Albuquerque: University of New Mexico Press, 1992.

Podell, Janet, and Seven Anzorin, eds. *Speeches of the American Presidents.* New York: H. W. Wilson, 1988.

Pollan, Michael. *Second Nature: A Gardener's Education.* New York: Bantam Doubleday, 1991.

Pomeroy, Earl. *The Pacific Slope: A History of California, Oregon, Washington, Idaho, Utah, and Nevada.* New York: Knopf, 1965.

Porter, Glenn, and Harold C. Livesay. *Merchants and Manufacturers: Studies in the Changing Structure of Nineteenth-Century Marketing.* Baltimore: Johns Hopkins Press, 1971.

Powell, G. Harold. *Coöperation in Agriculture.* New York: Macmillan, 1913.

———. *Letters from the Orange Empire.* Edited by Richard G. Lillard. Los Angeles: Historical Society of Southern California, 1990.

Preston, William L. *Vanishing Landscapes: Life and Land in the Tulare Lake Basin.* Berkeley and Los Angeles: University of California Press, 1981.

Rasmussen, Wayne D., ed. *Readings in the History of American Agriculture.* Urbana: University of Illinois Press, 1960.

Reisler, Mark. *By the Sweat of Their Brow: Mexican Immigrant Labor in the United States, 1900–1940.* West, Conn.: Greenwood Press, 1976.

Ricardo, David. *The Works and Correspondence of David Ricardo*, edited by Piero Sraffa, with the collaboration of M. H. Dobb. Vol. 1. Cambridge: Cambridge University Press, 1951; reprint, 1975.

Ridge, Martin. *Ignatius Donnelly: The Portrait of a Politician*. Chicago: University of Chicago Press, 1962.

Robinson, Jancis. *Vines, Grapes, and Wines*. New York: Knopf, 1986.

Rodgers, Andrew Denny III. *Liberty Hyde Bailey: A Story of American Plant Sciences*. Princeton, N.J.: Princeton University Press, 1949.

Roeding, George C. *Fruit Growers' Guide*. Fresno: George Roeding, 1919.

Rohrabacher, C. A. *Fresno County, California: Descriptive, Statistical and Biographical*. Fresno: C. A. Rohrabacher, 1891.

Ruiz, Vicki. *Cannery Women; Cannery Lives: Mexican Women, Unionization, and the California Food Processing Industry, 1930–1950*. Albuquerque: University of New Mexico Press, 1987.

Saker, Victoria Alice. "Benevolent Monopoly: The Legal Transformation of Agricultural Cooperation, 1890–1943." Ph.D. diss., University of California at Berkeley, 1990.

Sanderson, Dwight, ed. *Farm Income and Farm Life: A Symposium on the Relation of the Social and Economic Factors in Rural Progress*. Chicago: University of Chicago Press, 1927.

Saxton, Alexander. *The Indispensable Enemy: Labor and the Anti-Chinese Movement in California*. Berkeley and Los Angeles: University of California Press, 1971.

Schmidt, Louis Bernard, and Earle Dudley Ross, eds. *Readings in the Economic History of American Agriculture*. New York: Macmillan, 1925.

Seyd, Ernest. *California and Its Resources*. London: Trübner and Company, 1859.

Shannon, Fred A. *The Farmer's Last Frontier: Agriculture, 1860–1897*. New York: Farrar and Rinehart, 1945.

Shideler, James. *Farm Crisis: 1919–1923*. Berkeley and Los Angeles: University of California Press, 1957.

Shiva, Vandana. *Monocultures of the Mind: Perspectives on Biodiversity and Biotechnology*. London and New Jersey: Zed Books, 1993.

Smith, Joseph Russell. *North America: Its People and the Resources, Development, and Prospects of the Continent as an Agricultural, Industrial, and Commercial Area*. New York: Harcourt, Brace, and Company, 1925.

———. *Tree Crops: A Permanent Agriculture*. New York: Harcourt, Brace, and Company, 1929.

Smith, Wallace. *Garden of the Sun*. Fresno: A1 Printers, 1939.

Smythe, William. *The Conquest of Arid America*. London: Macmillan, 1905.

Snetsinger, Robert. *The Ratcatcher's Child*. Cleveland, Ohio: Franzak and Foster, 1983.

Snodgrass, Milton M., and Luther T. Wallace. *Agriculture, Economics, and Growth*. 2nd ed. New York: Appleton-Century-Croft, 1970.

Stadtman, Verne A. *The University of California, 1868–1968*. New York: McGraw-Hill Book Company, 1970.

Star, Kevin. *Inventing the Dream: California Through the Progressive Era*. New York: Oxford University Press, 1985.

Steen, Herman. *Coöperative Marketing: The Golden Rule in Agriculture*. Garden City, N.Y.: Doubleday, Page, and Company, 1923.

Stegner, Wallace. *Beyond the Hundredth Meridian: John Wesley Powell and the Second Opening of the West*. Lincoln: University of Nebraska Press, 1982.

Steinbeck, John. *The Grapes of Wrath*. Viking Press, 1939; reprint, New York: Penguin Books, 1985.

Symons, Leslie. *Agricultural Geography*. New York: Frederick A. Praeger, 1967.

Taylor, John. *Arator*. Petersburg, Virginia, 1818.

Taylor, Paul S. *Patterns of Agricultural Labor Migration within California*. Serial No. 840, United States Department of Labor, Bureau of Labor Statistics. Washington, D.C.: Government Printing Office, 1938.

Teague, Charles Collins. *Fifty Years a Rancher: The Recollections of Half a Century Devoted to the Citrus and Walnut Industries of California and to Furthering the Cooperative Movement in Agriculture*. [Los Angeles]: Charles Collins Teague, 1944.

Teele, Ray Palmer. *Irrigation in the United States*. New York: Appleton and Company, 1915.

Thickens, Virginia. *Pioneer Colonies of Fresno County*. Ph.D. diss., University of California at Berkeley, 1942.

Thornton, Tamara Plakins. *Cultivating Gentlemen: The Meaning of Country Life among the Boston Elite, 1785–1860*. New Haven: Yale University Press, 1989.

Trachtenberg, Alan. *The Incorporation of America: Culture and Society in the Gilded Age*. New York: Hill and Wang, 1982.

True, A. C. *A History of Agricultural Extension Work in the United States, 1785–1923*. Washington, D.C.: United States Department of Agriculture Miscellanies Publication 15, 1929.

True, A. C., and V. A. Clark. *The Agricultural Experiment Stations in the United States*. Washington: Government Printing Office, 1900.

Truettner, William H., ed. *The West as America: Reinterpreting Images of the Frontier, 1820–1920*. Washington, D.C.: Smithsonian Institution Press, 1991.

Turner, Frederick Jackson. *The Significance of Sections in American History*. New York: Henry Holt and Company, 1932.

Vandor, Paul E. *History of Fresno County, California, with Biographical Sketches*. 2 vols. Los Angeles: Historic Record Company, 1919.

Von Thünen, Johann Heinrich, *Der Isolierte Staat*. In *Von Thünen's Isolated State: An English Translation of Der Isolierte Staat*, ed. by Peter Hall and trans. by Carla M. Wartenberg. Oxford, England: Pergamon, 1966.

Warren, G. F. *Farm Management*. The Rural Text Book Series, ed. by L. H. Bailey. New York: Macmillan, 1919.

Webb, Walter Prescott. *The Great Plains*. 1931. Reprint, Lincoln: University of Nebraska Press, 1981.

Weber, Gustavus A. *The Plant Quarantine and Control Administration: Its History, Activities, and Organization*. Washington, D.C.: Brookings Institution, 1930.

White, Richard. *"It's Your Misfortune and None of My Own": A History of the American West*. Norman, Okla.: University of Oklahoma Press, 1991.

Whorton, James. *Before Silent Spring: Pesticides and Public Health in Pre-DDT America*. Princeton, N.J.: Princeton University Press.

Wickson, Edward James. *The California Fruits and How to Grow Them: A Manual of Methods Which Have Yielded Greatest Success, with the List of Varieties Best Adapted to the Different Districts of the State*. 6th ed. San Francisco: Pacific Rural Press, 1912.

———. *Rural California*. New York: Macmillan, 1923.

Willcox, O. W. *Reshaping Agriculture*. New York: W. W. Norton and Company, 1934.

Williams, Raymond. *The Country and the City*. New York: Oxford University Press, 1973.

Winchell, Lilbourne Alsip. *The History of Fresno County and the San Joaquin Valley*. Fresno: A. H. Cawston, 1933.

Winkler, Albert Julius. *General Viticulture*. Berkeley and Los Angeles: University of California Press, 1974.

Worster, Donald. *Dust Bowl: The Southern Plains in the 1930s*. Oxford: Oxford University Press, 1979.

———. *Rivers of Empire: Water, Aridity, and the Growth of the American West*. New York: Pantheon Books, 1985.

Wrobel, David M. *The End of American Exceptionalism: Frontier Anxiety from the Old West to the New Deal*. Lawrence: University Press of Kansas, 1993.

Wyckoff, William. *The Developer's Frontier: The Making of the Western New York Landscape*. New Haven: Yale University Press, 1988.

Zonlight, Margaret Aseman Cooper. *Land, Water, and Settlement in Kern County, California*. New York: Armo Press, 1979.

Index

Compositor:	G & S Typesetters, Inc.
Text:	10/13 Galliard
Display:	Galliard
Printer and Binder:	Thomson-Shore, Inc.